Information Theory Tools
for Image Processing

Synthesis Lectures on Computer Graphics and Animation

Editor

Brian A. Barsky, *University of California, Berkeley*

This series will present lectures on research and development in computer graphics and geometric modeling for an audience of professional developers, researchers and advanced students. Topics of interest include Animation, Visualization, Special Effects, Game design, Image techniques, Computational Geometry, Modeling, Rendering and others of interest to the graphics system developer or researcher.

Information Theory Tools for Image Processing

Miquel Feixas, Anton Bardera, Jaume Rigau, Qing Xu, and Mateu Sbert

ISBN: 978-3-031-79554-1   paperback
ISBN: 978-3-031-79555-8   ebook

DOI 10.1007/978-3-031-79555-8

A Publication in the Springer series
SYNTHESIS LECTURES ON COMPUTER GRAPHICS AND ANIMATION

Lecture #15
Series Editor: Brian A. Barsky, *University of California, Berkeley*
Series ISSN
Synthesis Lectures on Computer Graphics and Animation
Print 1933-8996   Electronic 1933-9003

Information Theory Tools
for Image Processing

Miquel Feixas
University of Girona, Spain

Anton Bardera
University of Girona, Spain

Jaume Rigau
University of Girona, Spain

Qing Xu
Tianjin University, China

Mateu Sbert
University of Girona, Spain

*SYNTHESIS LECTURES ON COMPUTER GRAPHICS AND ANIMATION
#15*

ABSTRACT

Information Theory (IT) tools, widely used in many scientific fields such as engineering, physics, genetics, neuroscience, and many others, are also useful transversal tools in image processing. In this book, we present the basic concepts of IT and how they have been used in the image processing areas of registration, segmentation, video processing, and computational aesthetics. Some of the approaches presented, such as the application of mutual information to registration, are the state of the art in the field. All techniques presented in this book have been previously published in peer-reviewed conference proceedings or international journals. We have stressed here their common aspects, and presented them in an unified way, so to make clear to the reader which problems IT tools can help to solve, which specific tools to use, and how to apply them. The IT basics are presented so as to be self-contained in the book. The intended audiences are students and practitioners of image processing and related areas such as computer graphics and visualization. In addition, students and practitioners of IT will be interested in knowing about these applications.

KEYWORDS

image registration, image segmentation, image fusion, video processing, computational aesthetics, information theory, entropy, mutual information, Jensen-Shannon divergence, information bottleneck method

Contents

Preface

Information theory (IT) tools are widely used in fields such as engineering, physics, genetics, neuroscience, and others, and have been also extensively applied to image processing. In this book, which follows a first book on IT applications to computer graphics and visualization [153], we present the basic concepts of IT and how they have been used in the image processing areas of registration, segmentation, video processing, and computational aesthetics. We deal in this book with the image as a map from a color palette to an array of pixels. We will not deal with how this map was obtained, that is, we don't deal with image formation. For applications of information theory to image formation see [119].

The intended audience are students and practitioners of image processing and related areas, as computer graphics. In addition, students and practitioners of IT will be interested in knowing about these applications. We believe that interest in this book can come out of several reasons. First, IT techniques are more and more pervading different areas, being more and more used as a technological tool. Thus, new applications of IT techniques will be considered with high interest. Second, image processing is being widely used in production. Soundly-based techniques will henceforth be most welcome. Third, medical imaging community is already using IT techniques as the most reliable ones. The growing importance of this community is influencing both related fields of computer graphics and visualization and image processing. Fourth point is taken from our own experience. In the courses and tutorials taught in several conferences, presentation of material of this book has risen so far a lot of interest and audience. In addition, some of the presented approaches, such as the use of mutual information in image registration, are the state of the art techniques, receiving more and more attention from practitioners.

This document is organized in the following way. After this preface, the first chapter presents the basics of IT. Information theory, introduced by Shannon in 1948, deals with the transmission, storage, and processing of information. Firstly, we introduce the concept of entropy, which can be thought of as a mathematical measure of information or uncertainty. Then, the measures of conditional entropy and mutual information are defined together with the concept of communication channel. Other measures, such as entropy rate, f-divergences, generalized entropies, and normalized compression distance, are also introduced. Finally, the information bottleneck method and the Jensen-Shannon inequality, which play an important role in some chapters of this book, are presented.

The second chapter deals with image registration. Image registration consists in finding the geometrical transform that aligns the input images into a unique coordinate space. It is of special interest in medical applications since it allows to integrate complementary information in a single image. In this chapter, we introduce the four elements of the image registration pipeline and several measures and techniques based on Shannon's information measures. The generalization of the information measures in order to include spatial information and the use of f-divergences for image registration are described. Other techniques that use generalized entropies to define new metrics or that use measures based on compression are also summarized. As a last point, an application of information-theoretic tools for image fusion is presented.

The third chapter is about image segmentation. The main goal of image segmentation is to subdivide an image into its constituent parts and is typically used to identify objects or other relevant information in digital images. Image segmentation is one of the most widely studied problems in image analysis and computer vision, and it is a significant step towards image understanding. In this chapter, the main proposed segmentation algorithms related to IT are summarized. First, several methods based on the maximum entropy criterion, which was the first attempt to introduce the information theory to image segmentation, and some extensions of these methods are introduced. Then, methods that introduce the spatial distribution of the intensity values in the image to obtain the optimal threshold values are presented. Next, different approaches that introduce IT measures in the energy functional of evolving curves for image segmentation are explained. Finally, methods based on the information bottleneck method applied to different information channels are presented.

The fourth chapter is focused on video processing techniques. Digital videos have become more and more popular, and applications require to have a good and quick understanding of general video sequences. Video key frames, which are a set of images representing the main content of the original video data, can be utilized for this purpose. As a result, key frame extraction is a fast and sound way to summarize a video sequence. The segmentation of a video sequence into its constituent shots, namely the shot boundary detection, is often a fundamental step for key frame selection. Thus, in this chapter, we present several efficient information theoretic techniques for video key frame selection and obtaining shot boundaries. The approaches are mainly based on mutual information and its extension to Tsallis mutual information, and the Jensen-Shannon divergence and its extensions to Jensen-Rényi divergence and Jensen-Tsallis divergence. These measures are investigated to estimate the frame-by-frame distance or similarity between consecutive video images, for recognizing shot or subshot boundaries and for choosing key frames.

The fifth chapter is devoted to computational aesthetics. This field was created in 1928 when George D. Birkhoff formalized the concept of aesthetic measure as the quotient between order and complexity. In this chapter, we focus our attention on informational aesthetics measures which have the goal of quantifying aesthetics. First, we present a set of measures that conceptualize the Birkhoff's aesthetic measure from an informational point of view. A first group of global measures, based on Shannon entropy and Kolmogorov complexity, gives us a scalar value associated with an artistic object. A second group of compositional measures extends the previous analysis in order to capture the structural information of the object. In particular, an information channel that fits well with the creative channel presented by Max Bense is introduced. These measures are applied to different sets of paintings, representing different styles, and to the study of the evolution of Van Gogh's artwork. Finally, we introduce two measures which quantify the information associated with each color and region of a painting. These measures permit us to visualize the most informative or salient colors and elements (objects or regions) of an image.

Miquel Feixas, Anton Bardera, Jaume Rigau, Qing Xu, and Mateu Sbert
March 2014

Acknowledgments

Authors wish to thank papers coauthors Imma Boada, Roger Bramon, Màrius Vila and Christian Wallraven for discussions and producing some of the images used in this book. We thank to third authors that allowed the use of their images in this book. Thanks also to David Brooks [22] for having allowed us the use of his images of Van Gogh's paintings. Special thanks go to Cosmin Ancuti for their most useful comments to an early draft of this book. Authors acknowledge support from grants TIN2010-21089-C03-01 of Spanish Government and 2009-SGR-643 of Generalitat de Catalunya (Catalan Government), and Qing Xu acknowledges support from Natural Science Foundation of China (61179067, 60879003).

Miquel Feixas, Anton Bardera, Jaume Rigau, Qing Xu, and Mateu Sbert
March 2014

CHAPTER 1

Information Theory Basics

In 1948, Claude Shannon published a paper entitled "A mathematical theory of communication" [157] which marks the beginning of information theory. In this paper, Shannon defined information measures such as entropy and mutual information,[1] and introduced the fundamental laws of data compression and transmission. Information theory deals with the transmission, storage, and processing of information and is used in fields such as physics, computer science, mathematics, statistics, economics, biology, linguistics, neurology, learning, computer graphics, and image processing.

In information theory, *information* is simply the outcome of a selection among a finite number of possibilities and an information source is modeled as a random variable or a random process. The classical measure of information, Shannon entropy, expresses the information content or the uncertainty of a single random variable. It is also a measure of the dispersion or diversity of a probability distribution of observed events. For two random variables, their mutual information is a measure of the dependence between them. Mutual information plays an important role in the study of a *communication channel*, a system in which the output depends probabilistically on its input [42, 177, 188].

This chapter presents Shannon's information measures (entropy, conditional entropy, and mutual information) and their most basic properties. Entropy rate, the information bottleneck method, generalized entropies, and the normalized compression distance are also introduced. Good references of information theory are the books by Cover and Thomas [42] and Yeung [188].

1.1 ENTROPY

After representing a discrete information source as a random process, Shannon asks himself: "Can we define a quantity which will measure, in some sense, how much information is produced by such a process, or better, at what rate information is produced?" [157].

In his answer, Shannon supposes that we have a set of possible events whose probabilities of occurrence are p_1, p_2, ..., p_n and asks for the possibility of finding a measure, denoted by $H(p_1, p_2, \ldots, p_n)$, of how much "choice" is involved in the selection of the event or of how uncertain we are of the outcome. If this uncertainty measure exists, it is reasonable to require of it the following properties.

1. H would be continuous in the p_i.

---

[1]In Shannon's paper, the mutual information is called rate of transmission.

2. If all the p_i are equal (i.e., $p_i = 1/n$), then H should be a monotonic increasing function of n. With equally likely events there is more choice, or uncertainty, when there are more possible events.

3. If a choice is broken down into two successive choices, the original H should be the weighted sum of the individual values of H. The meaning of this property, called grouping property, is illustrated in Figure 1.1.[2]

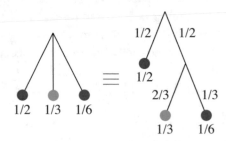

Figure 1.1: Grouping property of the entropy. On the left, we have three possibilities with probabilities $p_1 = 1/2$, $p_2 = 1/3$, $p_3 = 1/6$. On the right, we first choose between two possibilities each with probability 1/2, and if the second occurs, we make another choice with probabilities 2/3, 1/3. The final results have the same probabilities as before. In this example, it is required that $H(1/2, 1/3, 1/6) = H(1/2, 1/2) + (1/2)H(2/3, 1/3)$. The coefficient 1/2 is because the second choice occurs with this probability.

After these requirements, Shannon proved the following theorem.

Theorem 1.1 *The only measure H satisfying the three above assumptions is of the form*

$$H = -K \sum_{i=1}^{n} p_i \log p_i,$$ (1.1)

where K is a positive constant.[3]

To prove this theorem, Shannon assumed $H(1/n, 1/n, \ldots, 1/n) = f(n)$ and decomposed a choice from s^m equally likely possibilities into a series of m choices from s equally likely possibilities. Thus, from the previous required property (3), $f(s^m) = mf(s)$. In essence, this expression contains the intuition that the uncertainty of m choices should be m times the uncertainty of only one choice. One function that fulfills this requirement is the logarithm function (see the complete proof in [157]).

---

[2]This example has been used by Shannon in [157].
[3]This constant will be taken equal to 1 in the definition of entropy (Equation 1.2).

There are other axiomatic formulations which involve the same definition of uncertainty [42, 122]. Shannon called this quantity entropy,[4] as it can be identified with the entropy used in thermodynamics and statistical mechanics.

Let X be a discrete random variable[5] with alphabet $\mathcal{X}$ and probability distribution $\{p(x)\}$, where $p(x) = \Pr\{X = x\}$ and $x \in \mathcal{X}$. In this book, $\{p(x)\}$ will be also denoted by $p(X)$ or simply p. This notation will be extended to two or more random variables. As an example, a discrete random variable can be used to describe the toss of a fair coin, with alphabet $\mathcal{X} = \{head, tail\}$ and probability distribution $p(X) = \{1/2, 1/2\}$.

Definition 1.2   The entropy $H(X)$ of a discrete random variable X is defined by

$$H(X) = -\sum_{x \in \mathcal{X}} p(x) \log p(x), \tag{1.2}$$

where the summation is over the corresponding alphabet and the convention $0 \log 0 = 0$ is taken.

In this book, logarithms are taken in base 2 and, as a consequence, entropy is expressed in bits. The convention $0 \log 0 = 0$ is justified[6] by continuity since $x \log x \rightarrow 0$ as $x \rightarrow 0$. The term $-\log p(x)$ represents the information content (or uncertainty) associated with the result x. Thus, the entropy gives us the average amount of information (or uncertainty) of a random variable. Information and uncertainty are opposite. Uncertainty is considered before the event, information after. So, information reduces uncertainty. Note that the entropy depends only on the probabilities. We will use interchangeably the notation $H(X)$ or $H(p)$ for the entropy, where p stands for the probability distribution $p(X)$.

For example, the entropy of a fair coin toss is $H(X) = -(1/2) \log(1/2) - (1/2) \log(1/2) = \log 2 = 1$ bit. For the toss of a fair die with alphabet $\mathcal{X} = \{1, 2, 3, 4, 5, 6\}$ and probability distribution $p(X) = \{1/6, 1/6, 1/6, 1/6, 1/6, 1/6\}$, the entropy is $H(X) = \log 6 = 2.58$ bits.

Some relevant properties of the entropy [157] are

- $0 \leq H(X) \leq \log |\mathcal{X}|$

    - $H(X) = 0$ when all the probabilities are zero except one with unit value.
    - $H(X) = \log |\mathcal{X}|$ when all the probabilities are equal.

- If the probabilities are equalized, entropy increases.

---

[4]In the 18th century, R. Clausius introduced the term entropy in thermodynamics and L. Boltzmann gave its probabilistic interpretation in the context of statistical mechanics. The relationship between the Boltzmann entropy and Shannon entropy was developed in a series of papers by E. Jaynes [76]. The link between the second law of thermodynamics ("the entropy of an isolated system is non-decreasing") and the Shannon entropy is analyzed in [42].
[5]We assume that all random variables used are discrete unless otherwise specified.
[6]See in Yeung's book [188] for the discussion on probability distributions which are not strictly positive.

The binary entropy (Figure 1.2) of a random variable X with alphabet $\{x_1, x_2\}$ and probability distribution $\{p, 1 - p\}$ is given by

$$H(X) = -p \log p - (1 - p) \log(1 - p). \tag{1.3}$$

Note that the maximum entropy is $H(X) = 1$ bit when $p = 1/2$.

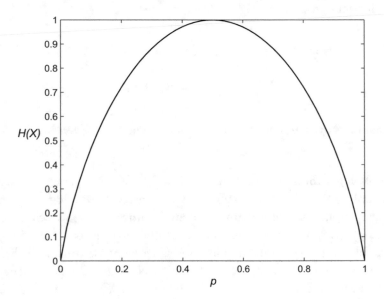

Figure 1.2: Plot of binary entropy.

The definition of entropy is now extended to a pair of random variables.

Definition 1.3   The joint entropy $H(X, Y)$ of a pair of discrete random variables X and Y with a joint probability distribution $p(X, Y) = \{p(x, y)\}$ is defined by

$$H(X, Y) = -\sum_{x \in \mathcal{X}} \sum_{y \in \mathcal{Y}} p(x, y) \log p(x, y), \tag{1.4}$$

where $p(x, y) = \Pr[X = x, Y = y]$ is the joint probability of x and y.

The conditional entropy of a random variable given another is defined as the expected value of the entropies of the conditional distributions.

Definition 1.4 The conditional entropy $H(Y|X)$ of a random variable Y given a random variable X is defined by

$$
\begin{aligned}
H(Y|X) &= \sum_{x \in \mathcal{X}} p(x) H(Y|X = x) = \sum_{x \in \mathcal{X}} p(x) \left(-\sum_{y \in \mathcal{Y}} p(y|x) \log p(y|x) \right) \\
&= -\sum_{x \in \mathcal{X}} \sum_{y \in \mathcal{Y}} p(x, y) \log p(y|x),
\end{aligned}
\tag{1.5}
$$

where $p(y|x) = \Pr[Y = y|X = x]$ is the conditional probability of y given x.[7]

The conditional entropy can be thought of in terms of a communication or *information channel* $X \to Y$ whose output Y depends probabilistically on its input X. This information channel is characterized by a transition probability matrix which determines the conditional distribution of the output given the input [42]. Hence, $H(Y|X)$ corresponds to the uncertainty in the channel output from the sender's point of view, and vice versa for $H(X|Y)$. Note that, in general, $H(Y|X) \neq H(X|Y)$. In this book, the conditional probability distribution of Y given x will be denoted by $p(Y|x)$ and the transition probability matrix (i.e., the matrix whose rows are given by $p(Y|x)$) will be denoted by $p(Y|X)$.

The following properties hold:

- $H(X, Y) = H(X) + H(Y|X) = H(Y) + H(X|Y)$

- $H(X, Y) \leq H(X) + H(Y)$

- $H(X) \geq H(X|Y) \geq 0$

- If X and Y are independent, then $H(Y|X) = H(Y)$ since $p(y|x) = p(y)$ and, consequently, $H(X, Y) = H(X) + H(Y)$ (i.e., entropy is additive for independent random variables).

As an example, we consider the joint distribution $p(X, Y)$ represented in Figure 1.3 *left*. The marginal probability distributions of X and Y are given by $p(X) = \{0.25, 0.25, 0.5\}$ and $p(Y) = \{0.375, 0.625\}$, respectively. Thus, $H(X) = -0.25 \log 0.25 - 0.25 \log 0.25 - 0.5 \log 0.5 = 1.5$ bits, $H(Y) = -0.375 \log 0.375 - 0.625 \log 0.625 = 0.954$ bits, and $H(X, Y) = -0.125 \log 0.125 - 0.125 \log 0.125 - 0.25 \log 0.25 - 0 \log 0 - 0 \log 0 - 0.5 \log 0.5 = 1.75$ bits.

---

[7]The Bayes theorem relates marginal probabilities $p(x)$ and $p(y)$, conditional probabilities $p(y|x)$ and $p(x|y)$, and joint probabilities $p(x, y)$:

$$
p(x, y) = p(x)p(y|x) = p(y)p(x|y).
\tag{1.6}
$$

If X and Y are independent, then $p(x, y) = p(x)p(y)$. Marginal probabilities can be obtained from $p(x, y)$ by summation: $p(x) = \sum_{y \in \mathcal{Y}} p(x, y)$ and $p(y) = \sum_{x \in \mathcal{X}} p(x, y)$.

$p(X,Y)$	y		$p(X)$
	y_1	y_2	
x_1	0.125	0.125	0.25
$\mathcal{X}$ x_2	0.25	0	0.25
x_3	0	0.5	0.5
$p(Y)$	0.375	0.625	
$H(X,Y) = 1.75$			

$p(Y\|X)$	y		$H(Y\|x \in \mathcal{X})$
	y_1	y_2	
x_1	0.5	0.5	$H(Y\|x_1) = 1$
$\mathcal{X}$ x_2	1	0	$H(Y\|x_2) = 0$
x_3	0	1	$H(Y\|x_3) = 0$
$H(Y\|X) = 0.25$			

Figure 1.3: Example of joint, marginal, and conditional probability distributions of random variables X and Y. On the left, joint distribution $p(X,Y)$, marginal distributions $p(X)$ and $p(Y)$, and joint entropy $H(X,Y)$. On the right, transition probability matrix $p(Y|X)$ and conditional entropy $H(Y|X)$.

From the transition probability matrix $p(Y|X)$ represented in Figure 1.3 *right*, we can compute $H(Y|X)$ as follows:

$$
\begin{aligned}
H(Y|X) &= \sum_{i=1}^{3} p(x_i) H(Y|X = x_i) \\
&= 0.25 \, H(Y|X = x_1) + 0.25 \, H(Y|X = x_2) + 0.5 \, H(Y|X = x_3) \\
&= 0.25 \times 1 + 0.25 \times 0 + 0.5 \times 0 = 0.25 \text{ bits.}
\end{aligned}
$$

1.2 RELATIVE ENTROPY AND MUTUAL INFORMATION

We now introduce two new measures, relative entropy and mutual information, which quantify the distance between two probability distributions and the shared information between two random variables, respectively.

Definition 1.5 The relative entropy or Kullback-Leibler distance $D_{KL}(p,q)$ between two probability distributions p and q, that are defined over the alphabet $\mathcal{X}$, is defined by

$$
D_{KL}(p,q) = \sum_{x \in \mathcal{X}} p(x) \log \frac{p(x)}{q(x)}. \tag{1.7}
$$

The conventions that $0 \log(0/0) = 0$ and $a \log(a/0) = \infty$ if $a > 0$ are adopted. The relative entropy satisfies the divergence or information inequality

$$
D_{KL}(p,q) \geq 0, \tag{1.8}
$$

with equality if and only if $p = q$. The relative entropy is also called information divergence [45] or informational divergence [188], and it is not strictly a metric[8] since it is not symmetric and does not satisfy the triangle inequality.

Definition 1.6   The mutual information $I(X;Y)$ between two random variables X and Y is defined by

$$
\begin{aligned}
I(X;Y) &= H(X) - H(X|Y) = H(Y) - H(Y|X) && (1.9) \\
&= \sum_{x \in \mathcal{X}} \sum_{y \in \mathcal{Y}} p(x,y) \log \frac{p(x,y)}{p(x)p(y)} && (1.10) \\
&= \sum_{x \in \mathcal{X}} p(x) \sum_{y \in \mathcal{Y}} p(y|x) \log \frac{p(y|x)}{p(y)}. && (1.11)
\end{aligned}
$$

Mutual information (MI) represents the amount of information that one random variable, the input of the channel, contains about a second random variable, the output of the channel, and vice versa. That is, mutual information expresses how much the knowledge of Y decreases the uncertainty of X, and vice versa. $I(X;Y)$ is a measure of the shared information or dependence between X and Y. Thus, if X and Y are independent, then $I(X;Y) = 0$. Note that the mutual information can be expressed as the relative entropy between the joint distribution and the product of marginal distributions:

$$I(X;Y) = D_{KL}(p(X,Y), p(X)p(Y)). \tag{1.12}$$

Mutual information $I(X;Y)$ fulfills the following properties:

- $I(X;Y) \geq 0$ with equality if and only if X and Y are independent
- $I(X;Y) = I(Y;X)$
- $I(X;Y) = H(X) + H(Y) - H(X,Y)$
- $I(X;Y) \leq \min\{H(X), H(Y)\}$
- $I(X;X) = H(X)$.

The relationship between Shannon's information measures can be expressed by a Venn diagram, as shown in Figure 1.4.[9] The correspondence between Shannon's information measures and set theory is discussed in [188].

For the example presented in Figure 1.3, the mutual information can be easily computed: $I(X;Y) = H(Y) - H(Y|X) = 0.954 - 0.25 = 0.704$ bits.

---

[8]A metric between x and y is defined as a function $d(x,y)$ that fulfills the following properties: (1) non-negativity: $d(x,y) \geq 0$, (2) identity: $d(x,y) = 0$ if and only if $x = y$, (3) symmetry: $d(x,y) = d(y,x)$, and (4) triangle inequality: $d(x,y) + d(y,z) \geq d(x,z)$.

[9]The information diagram does not include the universal set as in a usual Venn diagram.

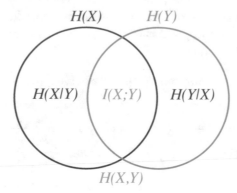

Figure 1.4: The information diagram represents the relationship between Shannon's information measures. Observe that $I(X;Y)$ and $H(X,Y)$ are represented, respectively, by the intersection and the union of the information in X (represented by $H(X)$) with the information in Y (represented by $H(Y)$). $H(X|Y)$ is represented by the difference between the information in X and the information in Y, and vice versa for $H(Y|X)$.

1.3   DECOMPOSITION OF MUTUAL INFORMATION

Given a communication channel $X \rightarrow Y$, mutual information can be decomposed in different ways to obtain the information associated with a value (or symbol) in $\mathcal{X}$ or $\mathcal{Y}$. Next, we present different definitions of information that have been proposed in the field of neural systems to investigate the significance associated to stimuli and responses, represented respectively by random variables S and R [24, 47].

For random variables S and R, representing an ensemble of stimuli $\mathcal{S}$ and a set of responses $\mathcal{R}$, respectively, mutual information (see Equations 1.9 and 1.11) is given by

$$I(S;R) \quad = \quad H(R) - H(R|S) \tag{1.13}$$

$$= \quad \sum_{s \in \mathcal{S}} p(s) \sum_{r \in \mathcal{R}} p(r|s) \log \frac{p(r|s)}{p(r)}, \tag{1.14}$$

where $p(r|s)$ is the conditional probability of value r known value s, and $p(S) = \{p(s)\}$ and $p(R) = \{p(r)\}$ are the marginal probability distributions of the input and output variables of the channel, respectively.

To quantify the information associated to each stimulus or response, $I(S;R)$ can be decomposed as

$$I(S;R) \quad = \quad \sum_{s \in \mathcal{S}} p(s) I(s;R) \tag{1.15}$$

$$= \quad \sum_{r \in \mathcal{R}} p(r) I(S;r), \tag{1.16}$$

where $I(s; R)$ and $I(S; r)$ represent, respectively, the information associated to stimulus s and response r. Thus, $I(S; R)$ can be seen as a weighted average over individual contributions from particular stimuli or particular responses. The definition of the contribution $I(s; R)$ or $I(S; r)$ can be performed in multiple ways, but we present here the three most basic definitions denoted by I_1, I_2, and I_3 [24, 47].

Given a stimulus s, three specific information measures that fulfill Equation 1.15 are defined.

Definition 1.7   The *surprise* I_1 can be directly derived from Equation 1.14, taking the contribution of a single stimulus to $I(S; R)$:

$$I_1(s; R) = \sum_{r \in \mathcal{R}} p(r|s) \log \frac{p(r|s)}{p(r)}. \tag{1.17}$$

The measure I_1 expresses the surprise about R from observing s. It can be shown that I_1 is the only positive decomposition of $I(S; R)$ [47]. This positivity can be proved from the fact that $I_1(s; R)$ is the Kullback-Leibler distance (Equation 1.7) between $p(R|s)$ and $p(R)$.

Definition 1.8   The *specific information* I_2 [47] can be derived from Equation 1.13, taking the contribution of a single stimulus to $I(S; R)$:

$$\begin{aligned} I_2(s; R) &= H(R) - H(R|s) \\ &= -\sum_{r \in \mathcal{R}} p(r) \log p(r) + \sum_{r \in \mathcal{R}} p(r|s) \log p(r|s). \end{aligned} \tag{1.18}$$

The measure I_2 expresses the change in uncertainty about R when s is observed. Note that I_2 can take negative values. This means that certain observations s do increase our uncertainty about the state of the variable R.

Definition 1.9   The *stimulus-specific information* I_3 is defined [24] by

$$I_3(s; R) = \sum_{r \in \mathcal{R}} p(r|s) I_2(S; r). \tag{1.19}$$

The measure I_3 also fulfills Equation 1.15 (for a proof, see [24]). A large value of $I_3(s; R)$ means that the states of R associated with s are very informative in the sense of $I_2(S; r)$. That is, the most informative input values s are those that are related to the most informative output values r.

Similar to the above definitions for a stimulus s, the information associated to a response r could be defined. The properties of positivity and additivity of these measures have been studied in [24, 47]. A measure is additive when the information obtained about X from two observations, $y \in \mathcal{Y}$ and $z \in \mathcal{Z}$, is equal to that obtained from y plus that obtained from z when y is known. While I_1 is always positive and non-additive, I_2 can take negative values but is additive, and I_3 can take negative values and is non-additive. On the one hand, because of the additivity property, DeWeese and Meister [47] prefer I_2 against I_1 since they consider that additivity is a fundamental property of any information measure. On the other hand, Butts [24] proposes some examples that show how I_3 identifies the most significant stimuli.

1.4   INEQUALITIES

In this section, we introduce a group of inequalities that are essential in the study of information theory and for the development of the concepts presented in this book [42, 188], and, in particular, to derive most of the refinement criteria.

1.4.1   JENSEN'S INEQUALITY

In this section, we introduce the concepts of convexity and concavity. Many important inequalities and results in information theory are obtained from the concavity of the logarithmic function.

Definition 1.10   A function $f(x)$ is convex over an interval $[a, b]$ (the graph of the function lies below any chord) if for every $x_1, x_2 \in [a, b]$ and $0 \leq \lambda \leq 1$,

$$f(\lambda x_1 + (1 - \lambda)x_2) \leq \lambda f(x_1) + (1 - \lambda)f(x_2). \tag{1.20}$$

A function is strictly convex if equality holds only if $\lambda = 0$ or $\lambda = 1$.

Definition 1.11   A function $f(x)$ is concave (the graph of the function lies above any chord) if $-f(x)$ is convex.

For instance, x^2 and $x \log x$ (for $x > 0$) are strictly convex functions, and $\log x$ (for $x > 0$) is a strictly concave function. Figure 1.5 plots $x \log x$ and $\log x$.

Jensen's inequality can be expressed as follows. If f is a convex function on the interval $[a, b]$, then

$$\sum_{i=1}^{n} \lambda_i f(x_i) - f\left(\sum_{i=1}^{n} \lambda_i x_i\right) \geq 0, \tag{1.21}$$

where $0 \leq \lambda \leq 1$, $\sum_{i=1}^{n} \lambda_i = 1$, and $x_i \in [a, b]$. If f is a concave function, the inequality is reversed. A special case of this inequality is when $\lambda_i = 1/n$ because then

$$\frac{1}{n} \sum_{i=1}^{n} f(x_i) - f\left(\frac{1}{n} \sum_{i=1}^{n} x_i\right) \geq 0, \tag{1.22}$$

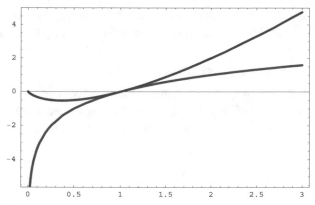

Figure 1.5: Plots of the strictly convex function $x \log x$ (red) and the strictly concave function $\log x$ (blue) for $x \in (0, 3]$.

that is, the value of the function at the mean of the x_i is less or equal than the mean of the values of the function at each x_i.

Jensen's inequality can also be expressed in the following way: if f is convex on the range of a random variable X, then

$$f(E[X]) \leq E[f(X)], \tag{1.23}$$

where E denotes expectation (i.e., $E[f(X)] = \sum_{x \in \mathcal{X}} p(x) f(x)$). Observe that if $f(x) = x^2$ (convex function), then $E[X^2] - (E[X])^2 \geq 0$. Thus, the variance is always positive.

One of the most important consequences of Jensen's inequality is the divergence inequality $D_{KL}(p, q) \geq 0$ (Equation 1.8). Some properties of Shannon's information measures presented in Sections 1.1 and 1.2 can be derived from this inequality.

1.4.2 LOG-SUM INEQUALITY

The log-sum inequality can be obtained from Jensen's inequality (Equation 1.21). For non-negative numbers $a_1, a_2, \ldots, a_n$ and $b_1, b_2, \ldots, b_n$, the log-sum inequality is expressed as

$$\sum_{i=1}^{n} a_i \log \frac{a_i}{b_i} - \left(\sum_{i=1}^{n} a_i \right) \log \frac{\sum_{i=1}^{n} a_i}{\sum_{i=1}^{n} b_i} \geq 0, \tag{1.24}$$

with equality if and only if a_i/b_i is constant for all i. The conventions that $0 \log 0 = 0$, $0 \log(0/0) = 0$, and $a \log(a/0) = \infty$ if $a > 0$ are again adopted.

From this inequality, the following properties can be proved [42].

- $D_{KL}(p, q)$ is convex in the pair (p, q).

- $H(X)$ is a concave function of p.

– If X and Y have the joint distribution $p(x, y) = p(x)p(y|x)$, then $I(X; Y)$ is a concave function of $p(x)$ for fixed $p(y|x)$ and a convex function of $p(y|x)$ for fixed $p(x)$.

1.4.3 JENSEN-SHANNON INEQUALITY

The Jensen-Shannon divergence, derived from the concavity of entropy, is used to measure the dissimilarity between two probability distributions and has the important feature that a different weight can be assigned to each probability distribution.

Definition 1.12   The Jensen-Shannon (JS) divergence is defined by

$$JS(\pi_1, \pi_2, \ldots, \pi_n; p_1, p_2, \ldots, p_n) = H\left(\sum_{i=1}^{n} \pi_i p_i\right) - \sum_{i=1}^{n} \pi_i H(p_i), \tag{1.25}$$

where $p_1, p_2, \ldots, p_n$ are a set of probability distributions defined over the same alphabet with prior probabilities or weights $\pi_1, \pi_2, \ldots, \pi_n$, fulfilling $\sum_{i=1}^{n} \pi_i = 1$, and $\sum_{i=1}^{n} \pi_i p_i$ is the probability distribution obtained from the weighted sum of the probability distributions $p_1, p_2, \ldots, p_n$.

From the concavity of entropy (Section 1.4.2), the Jensen-Shannon inequality [23] is obtained:

$$JS(\pi_1, \pi_2, \ldots, \pi_n; p_1, p_2, \ldots, p_n) \geq 0. \tag{1.26}$$

The JS-divergence measures how far the probabilities p_i are from their mixing distribution $\sum_{i=1}^{n} \pi_i p_i$, and equals zero if and only if all the p_i are equal. It is important to note that the JS-divergence is identical to the mutual information $I(X; Y)$ when $\pi_i = p(x_i)$ (i.e., $\{\pi_i\}$ corresponds to the marginal distribution $p(X)$), $p_i = p(Y|x_i)$ for all $x_i \in \mathcal{X}$ (i.e., p_i corresponds to the conditional distribution of Y given x_i), and $n = |\mathcal{X}|$ [23, 161].

1.4.4 DATA PROCESSING INEQUALITY

The data processing inequality is expressed as follows. If $X \rightarrow Y \rightarrow Z$ is a Markov chain,[10] then

$$I(X; Y) \geq I(X; Z). \tag{1.27}$$

This result proves that no processing of Y, deterministic or random, can increase the information that Y contains about X. In particular, if $Z = f(Y)$, then $X \rightarrow Y \rightarrow f(Y)$ and, consequently, $I(X; Y) \geq I(X; f(Y))$ [42].

---

[10]For random variables X, Y, and Z, $X \rightarrow Y \rightarrow Z$ forms a Markov chain if $p(x, y, z) = p(x)p(y|x)p(z|y)$. That is, the probability of the future state depends on the current state only and is independent of what happened before the current state. See a more general definition of Markov chain in Section 1.5.

1.5   ENTROPY RATE

Using the property $H(X_1, X_2) = H(X_1) + H(X_2|X_1)$ (Section 1.1) and the induction on n [188], it can be proved that the joint entropy of a collection of n random variables $X_1, \ldots, X_n$ is given by

$$H(X_1, \ldots, X_n) = \sum_{i=1}^{n} H(X_i|X_1, \ldots, X_{i-1}). \tag{1.28}$$

We now introduce the entropy rate that quantifies how the entropy of a sequence of n random variables increases with n.

Definition 1.13   The entropy rate or entropy density h^x of a stochastic process[11] $\{X_i\}$ is defined by

$$h^x \quad = \quad \lim_{n \to \infty} \frac{1}{n} H(X_1, X_2, \ldots, X_n) \tag{1.29}$$

when the limit exists.

The entropy rate represents the average information content per symbol in a stochastic process. For a stationary stochastic process,[12] the entropy rate exists and is equal to

$$h^x \quad = \quad \lim_{n \to \infty} h^x(n), \tag{1.30}$$

where $h^x(n) = H(X_1, \ldots, X_n) - H(X_1, \ldots, X_{n-1}) = H(X_n|X_{n-1}, \ldots, X_1)$. Entropy rate can be seen as the uncertainty associated with a given symbol if all the preceding symbols are known. The entropy rate of a sequence measures the average amount of information (i.e., irreducible randomness) per symbol x and the optimal achievement for any possible compression algorithm [42, 53].

An alternative notation, inspired by the work of Feldman and Crutchfield [43], is also used here to define the entropy rate. Given a chain $\ldots X_{-2} X_{-1} X_0 X_1 X_2 \ldots$ of random variables X_i taking values in $\mathcal{X}$, a block of L consecutive random variables is denoted by $X^L = X_1 \ldots X_L$. The probability that the particular L-block x^L occurs is denoted by $p(x^L)$. The joint entropy of length-L sequences or L-*block entropy* is now denoted by

$$H(X^L) = - \sum_{x^L \in \mathcal{X}^L} p(x^L) \log p(x^L), \tag{1.31}$$

---

[11]A stochastic process or a discrete-time information source $\{X_i\}$ is an indexed sequence of random variables characterized by the joint probability distribution $p(x_1, x_2, \ldots, x_n) = \Pr\{(X_1, X_2, \ldots, X_n) = (x_1, x_2, \ldots, x_n)\}$ with $(x_1, x_2, \ldots, x_n) \in \mathcal{X}^n$ for $n \geq 1$ [42, 188].

[12]A stochastic process $\{X_i\}$ is stationary if two subsets of the sequence, $\{X_1, X_2, \ldots, X_n\}$ and $\{X_{1+l}, X_{2+l}, \ldots, X_{n+l}\}$, have the same joint probability distribution for any $n, l \geq 1$: $\Pr\{(X_1, \ldots, X_n) = (x_1, x_2, \ldots, x_n)\} = \Pr\{(X_{1+l}, X_{2+l}, \ldots, X_{n+l}) = (x_1, x_2, \ldots, x_n)\}$. That is, the statistical properties of the process are invariant to a shift in time. At least, h^x exists for all stationary stochastic processes.

where the sum runs over all possible L-blocks. Thus, the *entropy rate* can be rewritten as

$$h^x = \lim_{L\to\infty} \frac{H(X^L)}{L} = \lim_{L\to\infty} h^x(L), \tag{1.32}$$

where $h^x(L) = H(X_L|X_{L-1}, X_{L-2}, \ldots, X_1)$ is the entropy of a symbol conditioned on a block of $L-1$ adjacent symbols.

A complementary measure to the entropy rate is the *excess entropy*, which is a measure of the *structural complexity* of a system. The *excess entropy* is defined by

$$E = \sum_{L=1}^{\infty} (h^x(L) - h^x) \tag{1.33}$$

$$= \lim_{L\to\infty} (H(X^L) - h^x L) \tag{1.34}$$

and captures how $h^x(L)$ converges to its asymptotic value h^x. Thus, when we take into account only a few number of symbols in the entropy computation, the system appears more random than it actually is. This excess of randomness tells us how much additional information must be gained about the configurations in order to reveal the actual uncertainty h^x. The way in which $h^x(L)$ converges to its asymptotic form tells us about the structure or correlations of a system [43, 54].

1.6   ENTROPY AND CODING

In this section, we present different interpretations of the Shannon entropy.

- As we have seen in Section 1.1, $-\log p(x)$ represents the information associated with the result x. The value $-\log p(x)$ can also be interpreted as the surprise associated with the outcome x. If $p(x)$ is small, the surprise is large; if $p(x)$ is large, the surprise is small. Thus, entropy (Equation 1.2) can be seen as the expectation value of the surprise [53].

- A fundamental result of information theory is the Shannon source coding theorem, which deals with the encoding of information in order to store or transmit it efficiently. This theorem can be formulated in the following ways [42, 53].

  - Given a random variable X, $H(X)$ fulfills

$$H(X) \leq \bar{\ell} < H(X) + 1, \tag{1.35}$$

  where $\bar{\ell}$ is the expected length of an optimal binary code for X. An example of an optimal binary code is the Huffman instantaneous coding.[13]

---

[13]A code is called a prefix or instantaneous code if no codeword is a prefix of any other codeword. Huffman coding uses a specific algorithm to obtain the representation for each symbol. The main characteristic of this code is that the most common symbols use shorter strings of bits than the ones used by the less common symbols.

- If we optimally encode n identically distributed random variables X with a binary code, the Shannon source coding theorem can be enunciated in the following way:

$$H(X) \leq \bar{\ell}_n < H(X) + \frac{1}{n}, \qquad (1.36)$$

where $\bar{\ell}_n$ is the expected codeword length per unit symbol. Thus, by using large block lengths, we can achieve an expected codelength per symbol arbitrarily close to the entropy [42].

- For a stationary stochastic process, we have

$$\frac{H(X_1, X_2, \ldots, X_n)}{n} \leq \bar{\ell}_n < \frac{H(X_1, X_2, \ldots, X_n)}{n} + 1 \qquad (1.37)$$

and, from the definition of entropy rate H_X (Equation 1.29),

$$\lim_{n \to \infty} \bar{\ell}_n \to H_X. \qquad (1.38)$$

Thus, the entropy rate is the expected number of bits per symbol required to describe the stochastic process.

- From the previous Shannon theorem, it can be proved that entropy is related to the difficulty in guessing the outcome of a random variable [42, 53] since

$$H(X) \leq \bar{q} < H(X) + 1, \qquad (1.39)$$

where $\bar{q}$ is the average minimum number of binary questions to determine X. This idea agrees with the interpretation of entropy as a measure of uncertainty.

1.7 CONTINUOUS CHANNEL

In this section, entropy and mutual information are defined for continuous random variables. Let X be a continuous random variable with continuous cumulative distribution function $F(x) = \Pr\{X \leq x\}$. When the derivative $F'(x) = f(x)$ is defined and $\int_{-\infty}^{\infty} f(x)\mathrm{d}x = 1$, then $f(x)$ is called the probability density function (pdf) of X. The support of X is given by $\mathcal{S}_X = \{x : f(x) > 0\}$, that is, the set of points where the function is non-zero. The statement "if it exists" should be included in the following definitions involving integrals and probability density functions. See a more detailed presentation in Cover and Thomas [42] and Yeung [188].

The differential entropy of a continuous random variable X is defined similarly to the entropy of a discrete random variable (see Equation 1.2).

Definition 1.14 The continuous or differential entropy $h(X)$ of a continuous random variable X with a pdf $f(x)$ is defined by

$$h(X) = -\int_{\mathcal{S}_X} f(x) \log f(x)\mathrm{d}x. \qquad (1.40)$$

Definition 1.15   For two continuous random variables X and Y with joint pdf $f(x, y)$, the continuous conditional entropy $h(Y|X)$ is defined as

$$h(Y|X) = -\int_{S_X} \int_{S_Y(x)} f(x, y) \log f(y|x) dx dy, \tag{1.41}$$

where $f(y|x)$ is the conditional pdf and $S_Y(x) = \{y : f(y|x) > 0\}$.

Definition 1.16   For two continuous random variables X and Y with joint pdf $f(x, y)$, the continuous mutual information $I^c(X; Y)$ is defined as

$$I^c(X; Y) = h(X) - h(X|Y) = \int_{S_X} \int_{S_Y(x)} f(x, y) \log \frac{f(x, y)}{f(x) f(y)} dx dy. \tag{1.42}$$

Following the exposition in [42], we divide the range of the continuous random variable X into discrete bins of length Δ (Figure 1.6 a). Then, assuming the continuity of $f(x)$ within the bins and using the mean value theorem, for each bin there exists a value x_i such that

$$f(x_i)\Delta = \int_{i\Delta}^{(i+1)\Delta} f(x) dx. \tag{1.43}$$

The discretized version of X is defined by

$$X_\Delta = x_i, \qquad \text{if } i\Delta \leq X < (i + 1)\Delta \tag{1.44}$$

with probability distribution $p(x_i) = \Pr\{X_\Delta = x_i\} = f(x_i)\Delta$. Thus, the entropy of X_Δ is given by

$$
\begin{aligned}
H(X_\Delta) &= -\sum_i p(x_i) \log p(x_i) = -\sum_i f(x_i)\Delta \log(f(x_i)\Delta) \\
&= -\sum_i f(x_i)\Delta \log f(x_i) - \log \Delta.
\end{aligned} \tag{1.45}
$$

If $f(x) \log f(x)$ is Riemann integrable, we obtain that

$$\lim_{\Delta \to 0} (H(X_\Delta) - \log \Delta) = h(X), \tag{1.46}$$

since $h(X) = \lim_{\Delta \to 0}(-\sum_i f(x_i)\Delta \log f(x_i))$. Thus, in general, the entropy of a continuous random variable does not equal the entropy of the discretized random variable in the limit of a

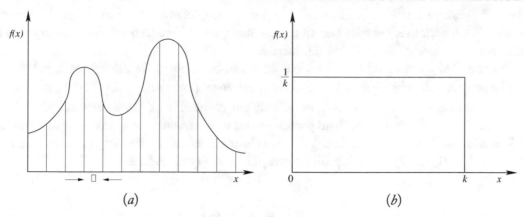

Figure 1.6: (a) Probability density function of X. The range of the continuous random variable X is divided into discrete bins of length Δ. (b) Constant probability density function $f(x) = 1/k$.

finer discretization. We can also see that, due to the fact that $-\lim_{\Delta \to 0} \log \Delta = \infty$, the entropy $H(X_\Delta)$ goes to infinity when the bin size goes to zero:

$$\lim_{\Delta \to 0} H(X_\Delta) = \infty. \tag{1.47}$$

For instance, if $f(x) = 1/k$ in the interval $(0, k)$ (Figure 1.6 b), then $h(X) = -\int_0^k (1/k) \log(1/k) \mathrm{d}x = \log k$. Observe that the differential entropy is negative when $k < 1$.

In contrast with the behavior of the differential entropy, the mutual information between two continuous random variables X and Y is the limit of the mutual information between their discretised versions. Thus, in the limit of a finer discretisation we get

$$I^c(X; Y) = \lim_{\Delta \to 0} I(X_\Delta; Y_\Delta). \tag{1.48}$$

Kolmogorov [85] and Pinsker [128] defined mutual information as $I^c(X; Y) = sup_{P,Q} I([X]_P; [Y]_Q)$, where the supremum ($sup$) is over all finite partitions P and Q of X and Y, respectively. From this definition and Equation 1.48, two important properties can be derived: the continuous mutual information is the least upper bound for the discrete mutual information and refinement can never decrease the discrete mutual information. This last property can also be deduced from the data processing inequality (Equation 1.27) [61].

1.8   INFORMATION BOTTLENECK METHOD

The information bottleneck method, introduced by Tishby et al. [169], is a technique that extracts a compact representation of the variable X, denoted by $\widehat{X}$, with minimal loss of mutual information with respect to another variable Y (i.e., $\widehat{X}$ preserves as much information as possible about the control variable Y). Thus, given an information channel between X and Y, the information bottleneck method tries to find the optimal tradeoff between accuracy and compression of X when the bins of this variable are clustered.

Soft [169] and hard [160] partitions of X can be adopted. In the first case, every $x \in \mathcal{X}$ can be assigned to a cluster $\hat{x} \in \widehat{\mathcal{X}}$ with some conditional probability $p(\hat{x}|x)$ (soft clustering). In the second case, every $x \in \mathcal{X}$ is assigned to only one cluster $\hat{x} \in \widehat{\mathcal{X}}$ (hard clustering).

In this book, we consider hard partitions and we focus our attention on the agglomerative information bottleneck method [160]. Given a cluster $\hat{x}$ defined by $\hat{x} = \{x_1, \ldots, x_l\}$, where $x_k \in \mathcal{X}$ for all $k \in \{1, \ldots, l\}$, and the probabilities $p(\hat{x})$ and $p(y|\hat{x})$ defined by

$$p(\hat{x}) = \sum_{k=1}^{l} p(x_k), \tag{1.49}$$

$$p(y|\hat{x}) = \frac{1}{p(\hat{x})} \sum_{k=1}^{l} p(x_k, y) \quad \forall y \in \mathcal{Y}, \tag{1.50}$$

the following properties are fulfilled.

– The decrease in the mutual information $I(X;Y)$ due to the merging of $x_1, \ldots, x_l$ is given by

$$\delta I_{\hat{x}} = p(\hat{x}) JS(\pi_1, \ldots, \pi_l; p_1, \ldots, p_l) \geq 0, \tag{1.51}$$

where the weights and probability distributions of the JS-divergence (Equation 1.25) are given by $\pi_k = p(x_k)/p(\hat{x})$ and $p_k = p(Y|x_k)$ for all $k \in \{1, \ldots, l\}$, respectively. An optimal clustering algorithm should minimize $\delta I_{\hat{x}}$.

– An optimal merging of l components can be obtained by $l - 1$ consecutive optimal mergings of pairs of components.

1.9 f-DIVERGENCES

Many different measures quantifying the divergence between two probability distributions have been studied in the past. They are frequently called "distances," although some of them are not strictly metrics. Some particular examples of divergences play an important role in different fields such as statistics and information theory [123].

Next, we present a measure of divergence between two probability distributions called *f*-divergence. This measure was independently introduced by Csiszár [44] and Ali and Silvey [5]. The following definition is taken from Csiszár and Shields [45].

Definition 1.17   Let $f(t)$ be a convex function defined for $t > 0$, with $f(1) = 0$. The *f*-divergence of a distribution p from q is defined by

$$D_f(p,q) = \sum_{x \in \mathcal{X}} q(x) f\left(\frac{p(x)}{q(x)}\right),\tag{1.52}$$

where the conventions $0f(0/0) = 0$, $f(0) = \lim_{t \to 0} f(t)$, $0f(a/0) = \lim_{t \to 0} t f(a/t) = a \lim_{u \to \infty}(f(u)/u)$ are adopted.

For the purposes of this book, we present three of the most important *f*-divergences: Kullback-Leibler, Chi-square, and Hellinger distances. These can be obtained from different convex functions f (see Figure 1.7):

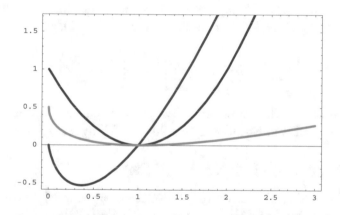

Figure 1.7: Plots for three strictly convex functions: $t \log t$ (blue), $(t-1)^2$ (red), and $1/2(1 - \sqrt{t})^2$ (green). From these functions, the Kullback-Leibler, Chi-square, and Hellinger distances are obtained, respectively.

Definition 1.18   Kullback-Leibler distance or information divergence [89]: If $f(t) = t \log t$, the Kullback-Leibler distance is given by

$$D_{KL}(p,q) = \sum_{x \in \mathcal{X}} p(x) \log \frac{p(x)}{q(x)}.\tag{1.53}$$

Definition 1.19 Chi-square distance [125]:

If $f(t) = (t-1)^2$, the Chi-square distance is given by

$$D_{\chi^2}(p,q) = \sum_{x \in \mathcal{X}} \frac{(p(x) - q(x))^2}{q(x)}. \tag{1.54}$$

Definition 1.20 Hellinger distance [67]:

If $f(t) = 1/2(1 - \sqrt{t})^2$, the Hellinger distance is given by

$$D_{h^2}(p,q) = \frac{1}{2} \sum_{x \in \mathcal{X}} (\sqrt{p(x)} - \sqrt{q(x)})^2. \tag{1.55}$$

Note that none of the above distances fulfills all the properties of a metric. However, the square root of the Hellinger distance is a true metric [50].

According to Csiszár and Shields [45], f-divergences generalize the Kullback-Leibler distance. Using the analog of the log-sum inequality (Section 1.4.2), given by

$$\sum_{i=1}^{n} b_i f\left(\frac{a_i}{b_i}\right) - \left(\sum_{i=1}^{n} b_i\right) f\left(\frac{\sum_{i=1}^{n} a_i}{\sum_{i=1}^{n} b_i}\right) \geq 0, \tag{1.56}$$

many of the properties of the information divergence extend to general f-divergences. If f is strictly convex the equality in Equation 1.56 holds if and only if a_i/b_i is constant for all i.

1.10 GENERALIZED ENTROPIES

Rényi [138] and Harvda and Charvát [65] introduced, respectively, two generalized definitions of entropy which includes the Shannon entropy as a particular case. Sharma and Mittal [158] and Sharma and Taneja [159] introduced two-parameter entropies where Rényi and Harvda-Charvát entropies are particular cases.

Definition 1.21 The *Rényi entropy* $H_\alpha^R(X)$ of a random variable X is defined by

$$H_\alpha^R(X) = \frac{1}{1-\alpha} \log \sum_{x \in \mathcal{X}} p(x)^\alpha, \tag{1.57}$$

where $\alpha > 0$ and $\alpha \neq 1$.

When $\alpha \to 1$, $H_\alpha^R(X) = H(X)$. $H_\alpha^R(X)$ is a concave function of p if $\alpha \leq 1$, but neither concave nor convex if $\alpha > 1$.

Tsallis [171] used the Harvda-Charvát entropy in order to generalize the Boltzmann entropy in statistical mechanics. The introduction of this entropy responds to the objective of generalizing the statistical mechanics to non-extensive systems.[14]

Definition 1.22   The Harvda-Charvát-Tsallis entropy $H_\alpha^T(X)$ of a discrete random variable X is defined by

$$H_\alpha^T(X) = k \frac{1 - \sum_{x \in \mathcal{X}} p(x)^\alpha}{\alpha - 1}, \tag{1.58}$$

where k is a positive constant (by default $k = 1$) and $\alpha \in \mathbb{R} \backslash \{1\}$ is called entropic index.

This entropy recovers the Shannon entropy (calculated with natural logarithms) when $\alpha \to 1$ and fulfils the properties of non-negativity and concavity (for $\alpha > 0$). In this book, the Harvda-Charvát-Tsallis entropy is also called Tsallis entropy.

If X and Y are independent, then the Harvda-Charvát-Tsallis entropy fulfills the non-additivity property:

$$H_\alpha^T(X, Y) = H_\alpha^T(X) + H_\alpha^T(Y) + (1 - \alpha) H_\alpha^T(X) H_\alpha^T(Y), \tag{1.59}$$

hence, superextensivity, extensivity, or subextensivity occurs when $\alpha < 1, \alpha = 1$ or $\alpha > 1$, respectively [173].

Definition 1.23   The *Tsallis conditional entropy* $H_\alpha(Y|X)$ is defined by

$$
\begin{aligned}
H_\alpha^T(Y|X) &= \sum_{x \in \mathcal{X}} p(x)^\alpha H_\alpha^T(Y|x) \\
&= \sum_{x \in \mathcal{X}} p(x)^\alpha \frac{1 - \sum_{y \in \mathcal{Y}} p(y|x)^\alpha}{\alpha - 1},
\end{aligned}
\tag{1.60}
$$

where $H_\alpha^T(Y|x)$ is the Tsallis entropy of Y known x.

Definition 1.24   Similar to Equation 1.10, the *Tsallis mutual information* $MI_\alpha^T(X;Y)$ is defined [167, 172] by

$$MI_\alpha^T(X;Y) = \frac{1}{1 - \alpha} \left(1 - \sum_{x \in \mathcal{X}} \sum_{y \in \mathcal{Y}} \frac{p(x, y)^\alpha}{p(x)^{\alpha-1} p(y)^{\alpha-1}} \right). \tag{1.61}$$

---

[14]An extensive system fulfills that quantities like energy and entropy are proportional to the system size. Similarly to Shannon entropy, a fundamental property of the Boltzmann entropy is its additivity. That is, if we consider a system composed by two probabilistically independent subsystems X and Y (i.e., $p(x, y) = p(x)p(y)$), then H(X,Y) = H(X) + H(Y). This property ensures the extensivity of the entropy but strongly correlated systems present non-extensive properties that require another type of entropy fulfilling non-additivity. Tsallis proposed the Harvda-Charvát entropy in order to deal with these "pathological" systems.

Definition 1.25   Another way of generalizing mutual information is the so-called *Tsallis mutual entropy* $ME_\alpha(X;Y)$, that, similar to Equation 1.9, is defined [57] by

$$
\begin{aligned}
ME_\alpha^T(X;Y) &= H_\alpha^T(Y) - H_\alpha^T(Y|X) \\
&= H_\alpha^T(X) + H_\alpha^T(Y) - H_\alpha^T(X,Y).
\end{aligned}
\tag{1.62}
$$

Furuichi [57] defined Tsallis mutual entropy for $\alpha > 1$ to ensure non-negativity, but for the purposes of this book this assumption is not necessary. Observe that both measures, $MI_\alpha^T(X;Y)$ and $ME_\alpha^T(X;Y)$, are different for $\alpha \neq 1$ and equal to the Shannon mutual information (calculated with natural logarithms) when $\alpha \to 1$.

1.11   THE SIMILARITY METRIC

The Kolmogorov complexity $K(x)$ of a string x is the length of the shortest program to compute x on an appropriate universal computer. Essentially, the Kolmogorov complexity of a string is the length of the ultimate compressed version of the string. The conditional complexity $K(x|y)$ of x relative to y is defined as the length of the shortest program to compute x given y as an auxiliary input to the computation. The joint complexity $K(x,y)$ represents the length of the shortest program for the pair (x,y) [95, 96].

The Kolmogorov information distance is defined as the length of the shortest program that computes x from y and y from x [15].

Definition 1.26   Up to an additive logarithmic term, the information distance is defined by

$$
E(x,y) = \max\{K(y|x), K(x|y)\}.
\tag{1.63}
$$

It can be shown that $E(x,y)$ is a metric [15]. It is interesting to note that long strings that differ by a tiny part are intuitively closer than short strings that differ by the same amount. Hence, there arises the necessity to normalize the information distance.

A normalized version of $E(x,y)$, called the normalized information distance or the similarity metric, was introduced by Li et al. [95].

Definition 1.27   The normalized information distance $NID(x,y)$ or the similarity metric is defined by

$$
\begin{aligned}
NID(x,y) &= \frac{\max\{K(x|y), K(y|x)\}}{\max\{K(x), K(y)\}} \\
&= \frac{K(x,y) - \min\{K(x), K(y)\}}{\max\{K(x), K(y)\}}.
\end{aligned}
\tag{1.64}
$$

$NID(x, y)$ is a metric and takes values in $[0, 1]$. It is also universal in the sense that if two strings are similar according to the particular feature described by a particular normalized admissible distance (not necessarily metric), then they are also similar in the sense of the normalized information metric.

Due to the non-computability of Kolmogorov complexity, a feasible version of $NID(x, y)$, called the normalized compression distance, was defined from the lengths of compressed data files [95].

Definition 1.28   The normalized compression distance $NCD(x, y)$ is defined by

$$NCD(x, y) = \frac{C(x, y) - \min\{C(x), C(y)\}}{\max\{C(x), C(y)\}}, \tag{1.65}$$

where $C(x)$ and $C(y)$ represent the lengths of compressed strings x and y, respectively, and $C(x, y)$ the length of the compressed pair (x, y).

Thus, NCD approximates NID by using a standard real-world compressor.

CHAPTER 2

Image Registration

Registration is a fundamental task in image processing used to match two or more images or datasets obtained at different times, from different devices or from different subjects. Basically, it consists in finding the geometrical transform that aligns these datasets into a unique coordinate space. In medical applications, image registration is of special interest since it allows us to integrate complementary information in a single dataset. The integration of information from different imaging modalities is difficult and, in most cases, dependent on the data we have to deal with. In Figure 2.1 two multimodal images ((a) Computerized Tomography (CT) and (b) Magnetic Resonance (MR) Imaging) are shown. Figures 2.1(c) and 2.1(d) show the fused images resulting from the direct combination (before the registration process) and from the combination after the registration process, respectively

The image registration pipeline starts with the selection of the two images to be registered. One of the two images is defined as the fixed image and the other one as the moving image. Given these inputs, image registration is treated as an optimization problem with the goal of finding the spatial mapping that brings the moving image into alignment with the fixed image. This process is composed of four basic elements [90, 129]: the transform, the interpolator, the metric, and the optimizer (see Figure 2.2). The *transform* represents the spatial mapping of points from the fixed image space to points in the moving image space. The *interpolator* is used to evaluate the moving image intensity values at non-grid positions. The *metric* provides a measure of how well the fixed image is matched by the transformed moving image. This measure is used as the quantitative

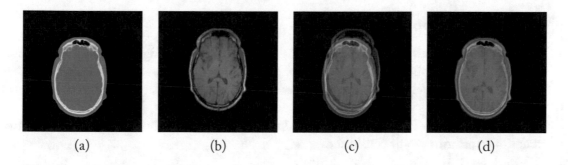

(a) (b) (c) (d)

Figure 2.1: (a) CT and (b) MR images, (c) the fused image at the original position, and (d) the fused image at the registration position.

criterion to be optimized by the *optimizer* over the search space defined by the parameters of the transform.

In this chapter, the image registration pipeline is introduced in more detail in Section 2.1. Then, different metrics based on Shannon's information measures and their computation are presented in Section 2.2, as well as different techniques to estimate these measures in Section 2.3. A generalization of the information measures in order to include spatial information is presented in Section 2.4. The use of f-divergences for image registration is described in Section 2.5 and the use of generalized entropies to define new metrics is presented in Section 2.6. Some measures based on compression and the normalized information distance are summarized in Section 2.7. Finally, the application of information-theoretic tools to image fusion is presented in Section 2.8.

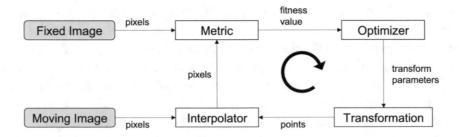

Figure 2.2: The main components of the image registration framework are the two input images, a transform, a metric, an interpolator, and an optimizer.

2.1   THE REGISTRATION PIPELINE

In this section the four main parts of the image registration pipeline: transform, interpolation, metric, and optimization are introduced. In Figure 2.2, the relationship between these parts is represented.

2.1.1   SPATIAL TRANSFORM

The registration process consists in reading the input datasets, defining the reference space (i.e., its resolution, positioning, and orientation) for each of these datasets, and establishing the *correspondence* between them (i.e., how to transform the coordinates of a dataset to the coordinates of the other dataset).

The spatial transform defines the spatial relationship between both images. Basically, two groups of transforms can be considered.

(a) *Rigid or affine transforms.* These transforms can be defined with a single global transform matrix. Rigid transforms are defined as geometrical transforms that only consider translations and rotations, and, thus, they preserve Euclidean distances between all the points.

Affine transforms also allow shearing transforms and they preserve the straightness of lines (and the planarity of surfaces) but not the distances. Affine transforms are mainly used in intrapatient registration (i.e., the registration of images of the same patient) or as a pre-processing step for nonrigid registration. The registration process between two 3D datasets requires the optimization of 6 parameters (3 for translation and 3 for rotation) when a rigid transform is used, while it requires 12 parameters when an affine transform is used.

(b) *Non-rigid or elastic transforms.* There are many different nonrigid transformation models that can be divided in two categories: physical-based models and function representations [73]. The first ones simulate some physical processes such as elasticity or fluid flow. The second ones arise from interpolation and approximation theory, and use basis functions to model the deformation. There are many different types of basis functions, e.g., radial basis functions, B-splines, and wavelets. Using these kinds of transforms, the straightness of the lines are not ensured. These transforms are very useful for interpatient registration (i.e., the registration between different subjects). The fusion of interpatient datasets allows the researchers to obtain the atlas of a given population, which permits to establish the general templates of diseases like schizophrenia or Alzheimer's.

2.1.2 INTERPOLATION

The interpolation strategy determines the intensity value of a point at a non-grid position. When a general transform is applied to an image, the transformed points may not coincide with the regular grid. So, an interpolation scheme is needed to estimate the values at these positions.

Several interpolation schemes have been introduced [93]. The most commonly used in the image registration domain are the following.

(a) *Trilinear interpolation.* The intensity of a point is obtained from the weighted combination of the intensities of its neighbors. The weights for the 2D case are plotted in Figure 2.3. Then, the intensity value at position $\mathcal{T}(x)$ is given by

$$f\left(\mathcal{T}(x)\right) = \sum_{i=1}^{4} \overline{w_i} f\left(n_i\right), \qquad (2.1)$$

where $\overline{w_i} = \frac{w_i}{\sum_{i=1}^{4} w_i}$.

(b) *Nearest neighbor interpolation.* The intensity of a point is the same as the nearest grid point.

(c) *Partial volume interpolation.* The weights of the linear interpolation (see Figure 2.3) are used to update the histogram bin of the corresponding grid intensity values [104]. In this way, no new intensity value pairs are introduced in the joint histogram.

(d) *Splines.* The intensity of a point is obtained from the spline-weighted combination of a grid-point kernel [175].

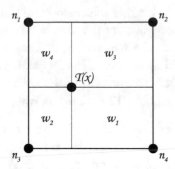

Figure 2.3: Interpolation weights, the areas w_i, for 2D linear interpolation.

All these strategies can introduce artifacts that deteriorate the accuracy and reliability of the registration. For the trilinear interpolation, these artifacts result from the low filter effect of the interpolator. For the partial volume interpolation, these artifacts are due to the higher histogram dispersion at non-grid positions [131]. The nearest neighbor interpolation causes artifacts due to the discontinuities that can appear for small transform differences.

Some works have dealt with the problem of reducing these artifacts. Tsao [174] showed that jittered sampling is extremely beneficial to the robustness and accuracy of registration, reducing considerably the interpolation artifacts. Since grid effects are caused by the regular grid sampling of the images, other stochastic sampling strategies have also been proposed [176]. Salvado and Wilson [150] applied a constant variance filter to the images to be registered in order to compensate the blurring effect of the linear interpolation.

2.1.3   METRIC

The metric evaluates the similarity (or disparity) between the two images to be registered. Several image similarity measures have been proposed, which can be classified depending on the features that they use.

(a) *Geometrical features.* A segmentation process detects in both images some image features, which are used for image registration. Measures based on geometric features minimize spatial disparity between selected features from the images (e.g., distance between corresponding points). These methods obtain subvoxel accuracy and they do not have high computational cost. Nevertheless, they have a great dependence on the initial segmentation results. The main difference between the methods within this group is the feature selected for the registration process: points, surfaces, intrinsic features such as landmarks, or extrinsic measures such as implanted markers.

(b) *Correlation measures.* In these cases, the alignment is achieved when a certain correlation measure between the intensity values of both images is maximized. Usually, a priori information is used in this kind of measures, since a known relationship (typically linear) between the intensity values of the datasets to be registered is assumed. This assumption is not valid for a general multimodal registration and, thus, most of the techniques of this group has been developed for monomodal registration. An important aspect to be considered is that correlation measures are only computed on the overlap area between both images, which varies according to the parameters of the transform.

(c) *Intensity occurrence.* These measures depend on the probability of each intensity value and on the joint histogram between both images. Information theory provides several measures to evaluate the disparity of the values of the joint histogram and, hence, the similarity between the images. Since the disparity of the joint histogram does not depend on the intensity values, most of the techniques of this group have been developed for multimodal registration [51, 132].

Despite this variety of measures, the latter group has become the most popular. In Section 2.2, a review of the main similarity metrics based on Shannon information measures is presented.

2.1.4 OPTIMIZATION

The optimizer finds the maximum (or minimum) value of the metric varying the parameters of the spatial transform. For the registration problem, an analytical solution is not possible. Then, numerical methods can be used in order to obtain the global extreme of a non analytical function. The most used methods in the medical image registration context are Powell's method, simplex method, gradient descent, conjugate-gradient method, quasi-Newton method, and evolutionary algorithms [84].

The choice of a method will depend on the implementation criteria and the measure features such as smoothness and robustness. A detailed description of several numerical optimization methods and their implementation can be found in [135]. Also, a study of the performance of some of these methods in image registration based on mutual information is done in [105].

2.2 SIMILARITY METRICS BASED ON SHANNON'S INFORMATION MEASURES

The registration metric quantifies the similarity (or disparity) between two images for a given transformation. It is considered that two datasets are registered when the similarity (or disparity) function is maximum (or minimum). The most successful automatic image registration methods are based on the mutual information (MI) maximization. This method, almost simultaneously introduced by Maes et al. [104] and Viola et al. [179], is based on the conjecture that the correct registration corresponds to the maximum MI between the overlapping areas of the two images. Later, Studholme et al. [164] proposed a normalization of MI which is more robust than MI, due to its greater independence of the overlap area.

In most of the information-theoretic methods, a crucial step is the computation of the joint histogram. The joint histogram depends on the transformation applied to the moving image B, since, for each transformation $\mathcal{T}(B)$, the overlapping area changes. When the images are correctly registered, there is an overlap between the corresponding anatomical structures and the joint histogram shows certain clusters for the intensity values of those structures. Conversely, when the images are misaligned, the structures in one image overlap with structures in the other image that are not their anatomical counterparts [132]. The basic idea behind these methods is that two values are related (or similar) when there are many other examples of those values occurring together in the overlapping image volume. These information-theoretic measures constitute a type of more generic statistical measures which only look at the occurrence of image values and not at the values themselves.

Figures 2.4(a) and 2.4(b) show a synthetic MR-T1 and MR-T2 image pair from the Brain-web database [38] and Figures 2.4(c-e) show the corresponding joint histograms at (c) the registration position and with a lateral translation of (d) 5 pixels and (e) 10 pixels. In these plots, the intensity at one point (i, j) is given by the number of pixels that have intensity i in the image 2.4(a) and intensity j in the image 2.4(b). This representation is done in the logarithm scale for a better visualization of the results. Note the decrease of the intensity of the clusters for corresponding anatomical structures and how new combinations of grey values emerge when the transform gets further away from the registration position. This fact is manifested in the joint histogram by an increase of the dispersion. This property is exploited by using measures of clustering or dispersion which have to be maximized or minimized, respectively. In this section, an overview of the main proposed measures based on Shannon information is presented.

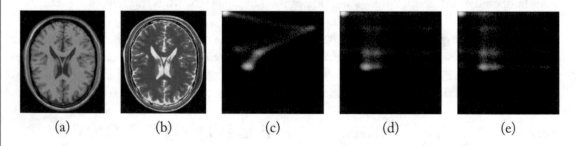

| (a) | (b) | (c) | (d) | (e) |

Figure 2.4:   Joint histogram plots for (a) MR-T1 and (b) MR-T2 test image pair at (c) the registration position and with a lateral translation of (d) 5 pixels, and (e) 10 pixels.

2.2.1   INFORMATION CHANNEL

In image registration, the information-theoretic measures have been often presented as statistical measures of the dependence between the values of both images. However, in some references (e.g., [8, 166]), the registration problem has been presented from an information channel. Here, we present this formulation as enables us to integrate all the measures in the same theoretical framework.

The registration process of two images is represented by an information channel $X \to Y$ (see Section 1.1), where the random variables X and Y represent, respectively, the images A and $\mathcal{T}$(B) (i.e., the image B after applying the transform $\mathcal{T}$). Their marginal probability distributions, $p(X)$ and $p(Y)$, and the joint probability distribution, $p(X, Y)$, are obtained by simple normalization of the marginal and joint intensity histograms of the overlapping areas of both images. The conditional probability distributions can be calculated using the Bayes theorem (see Equation 1.6), leading to the transition probability matrix $p(Y|X)$ of the channel (conditional probability matrix):

$$p(Y|X) = \begin{pmatrix} p_{y_1|x_1} & p_{y_2|x_1} & \cdots & p_{y_m|x_1} \\ p_{y_1|x_2} & p_{y_2|x_2} & \cdots & p_{y_m|x_2} \\ \vdots & \vdots & \ddots & \vdots \\ p_{y_1|x_n} & p_{y_2|x_n} & \cdots & p_{y_m|x_n} \end{pmatrix}, \tag{2.2}$$

where n and m are, respectively, the number of bins of the intensity histograms of images A and $\mathcal{T}$(B). The row x of this matrix is represented by the probability distribution $p(Y|x)$. For the inverse channel $Y \to X$, $p(X|Y)$ and $p(X|y)$ are, respectively, the conditional probability matrix and the row y of this matrix. The definition of this theoretical framework enriches the interpretation of the information-theoretic measures for registration. Thus, for instance, the maximization of the mutual information can be seen as the maximization of the channel capacity.

2.2.2   JOINT ENTROPY

The first attempts in the use of the joint histogram in order to compute the similarity between two images were addressed to compute its moments. This was first proposed by Hill [71] from visual examination of the effects of misregistration on the joint histogram, which peaks and valleys are more prominent at the registration position. The third moment or *skewness* was proposed in order to quantify this property. The skewness characterizes the degree of asymmetry of a distribution around its mean. It is a pure number that characterizes only the shape of the distribution [135].

Later, the information-theoretic measures were used as registration similarity measures. First, both Collignon et al. [41] and Studholme et al. [165] proposed the assumption that grey values disperse with misregistration and therefore the joint entropy can be used as a dissimilarity metric. By finding the transform that minimizes their joint entropy $H(X, Y)$ (see Equation 1.4), images should be registered. The main drawback of this method is its high sensitivity to the overlap area, since as this area decreases the amount of information also decreases. For instance, if the overlapping area contains only one voxel, the joint entropy is zero and, therefore, the minimum value is obtained, but obviously this dummy transform is not the one that moves to the correct registration position. Figure 2.5(a) shows a representation of $H(X, Y)$ over a translation from the registration position of a MR image (Figure 2.1(b)) to a CT one (Figure 2.1(a)).

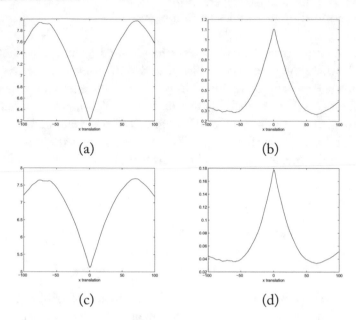

Figure 2.5:   Evaluation of some information measures of a translational movement of the CT and MR images of Figures 2.1(a) and 2.1(b). The evaluated measures are (a) $H(X,Y)$, (b) $I(X;Y)$, (c) $\rho(X,Y)$, and (d) $NMI(X,Y)$.

2.2.3   MUTUAL INFORMATION

One of the most used measures for multimodal image registration is the mutual information $I(X;Y)$ (see Equation 1.9). As it has been previously mentioned, mutual information quantifies the shared information between two random variables or the reduction of uncertainty of a random variable when the other variable is observed. Registration is assumed to correspond to the maximum mutual information: the images have to be aligned in such a manner that the amount of information they contain about each other is maximal [104, 184]. The more dependent the datasets, the higher the MI between them. Experiments have shown that MI is less sensitive to the overlap area than the joint entropy $H(X,Y)$. For instance, when the transform leads to a single voxel overlap, mutual information is 0 and, therefore, it is not the optimal value. Figure 2.5(b) shows a representation of $I(X;Y)$ over a translation of a MR image to a CT one.

Pluim et al. [132] presented an interesting dissertation about three different interpretations of the mutual information measure for image registration. The first one comes from the definition of mutual information (Equation 1.9) as

$$I(X;Y) \;\; = \;\; H(Y) - H(Y|X). \tag{2.3}$$

Here, the first term refers to the information content of the intensities of the image $\mathcal{T}(\mathsf{B})$, while the second refers to the information (or uncertainty) of the intensities of the image $\mathcal{T}(\mathsf{B})$ when the intensity of the image A at the corresponding point is known. Observe also that the second term decreases when the registration transform becomes closer to the registration position. The same reasoning can be applied to the other image, since Equation 2.3 is symmetric.

The second interpretation derives from the following expression:

$$I(X;Y) \quad = \quad H(X) + H(Y) - H(X,Y). \tag{2.4}$$

Note that this expression only depends on marginal and joint entropies, and not on conditional entropy. At first sight it seems that this measure should behave similarly than minus the joint entropy $-H(X,Y)$ (which was earlier proposed by Collignon et al. [41] and Studholme et al. [165]), since the other two terms correspond to the marginal entropies of both images which seem to be constant. But this conclusion is not true at all, since the marginal entropies vary for different transforms due to the fact that the area of overlap also changes. Remember that these measures are only computed from the samples inside the area of overlap. With these two terms (the marginal entropies), the mutual information measure becomes more robust than the joint histogram, reducing the sensitivity to the area of overlap. For instance, when evaluating the measure for an overlapping region containing only the background of both images, the joint entropy takes a very low value, and the mutual information also takes a very low value. Thus, while the minimization of the joint entropy would indicate that this is a good registration position (which is not desirable), the maximization of the mutual information would indicates that this is not (which is desirable).

The third interpretation comes from Equation 1.12:

$$
\begin{aligned}
I(X;Y) \quad &= \quad \sum_{x \in \mathcal{X}} \sum_{y \in \mathcal{Y}} p(x,y) \log \frac{p(x,y)}{p(x)p(y)} \\
&= \quad D_{KL}(p(X,Y), p(X)p(Y)).
\end{aligned}
\tag{2.5}
$$

This expression defines the mutual information as the Kullback-Leibler distance (see Equation 1.7) between the joint probability $p(X,Y)$ and the product of marginal probabilities $p(X)p(Y)$. It has to be taken into account that the product of marginal probabilities $p(X)p(Y)$ can be seen as the joint probability of the variables X and Y if they are independent. Therefore, MI can be seen as a measure of how far is the real joint probability from the independence distribution. In other words, mutual information is a measure of non-independence. Therefore, at the registration position, it is assumed to be as far as possible from the independence of both images.

Another measure, proposed by Maes et al. [104], related with mutual information is given by the *information distance*

$$
\begin{aligned}
\rho(X,Y) \quad &= \quad H(X,Y) - I(X;Y) \\
&= \quad H(X|Y) + H(Y|X).
\end{aligned}
\tag{2.6}
$$

It can be shown that $\rho(X,Y)$ is a true metric, since it is positive, symmetric, and fulfils the triangular inequality [42]. Maes et al. noted that, in some cases, this measure performs better than MI, but a clear preference for $\rho(X, Y)$ could not be established. Figure 2.5(c) shows a representation of $\rho(X, Y)$ over a translation of a MR image to a CT one.

2.2.4   NORMALIZED MEASURES

In the image registration context, Studholme [164] proposed a normalized measure of mutual information defined by

$$NMI(X, Y) = \frac{H(X) + H(Y)}{H(X, Y)} = 1 + \frac{I(X; Y)}{H(X, Y)}. \tag{2.7}$$

This measure takes values in the range $[1, 2]$. Nevertheless, in some works the normalized mutual information has been defined as $I(X; Y)/H(X, Y)$, since this ratio takes values in the most natural range $[0, 1]$. The necessity of normalization is theoretically justified by Li et al. [95]. Note that $1 - NMI$ is a true distance (see [87, 95]).

Experiments have shown that $NMI(X, Y)$ is more robust than $I(X; Y)$, due to its greater independence of the overlap area. This fact can be explained in the following manner. The size of the overlapping part of the images influences the mutual information measure in two ways. First of all, a decrease in the overlap area involves a decrease in the number of samples, which reduces the precision of the probability distribution estimation. Secondly, Studholme et al. [164, 166] showed that with an increasing misregistration (which usually coincides with a decreasing overlap) the mutual information value may actually increase. This can occur when the relative areas of object and background even out and the sum of the marginal entropies increases, faster than the joint entropy. The behavior of the normalized measure $NMI(X, Y)$ for the rigid registration of MR-CT and MR-PET images has been proved to be better than the non-normalized measures [164]. Figure 2.5(d) shows a representation of $NMI(X, Y)$ over a translation of a MR image to a CT one.

Collignon [40] and Maes [104] also suggested the use of the entropy correlation coefficient (ECC), another form of the normalized mutual information, which is defined as

$$
\begin{aligned}
ECC(X, Y) &= \frac{2I(X; Y)}{H(X) + H(Y)} \\
&= 2 - \frac{2}{NMI(X, Y)}.
\end{aligned} \tag{2.8}
$$

From this equation, it can be seen that the maximum of $NMI(X, Y)$ always coincides with the minimum of $ECC(X, Y)$ and, thus, these measures perform equally in the registration process.

2.3   PROBABILITY DENSITY FUNCTION ESTIMATION

For the computation of the information-theoretic measures presented in the previous section, each image is considered as a random variable source with an associated probability density function (pdf). Since we are interested on the relationship between two images, the joint probability of both random variables has to be estimated and, at this point, the transform and the interpolator have a critical role. First, the point correspondence is given by the transform and, second, the estimations of the intensity at the transformed points (usually at non-grid positions) are given by the interpolator.

In order to estimate these pdf's, two main schemes have been proposed: the histogram approach [104] and some non-parametric approaches, such as Parzen window method [124, 179] and the entropic spanning graph method [70].

2.3.1   HISTOGRAM ESTIMATION

A histogram is a function that counts the number of observations that fall into disjoint categories (known as bins) that represent the different values of a random variable. Histogram techniques compute the joint pdf by binning the intensity pairs $(A, \mathcal{T}(B))$ of the overlapping parts of the reference image A and the transformed image $\mathcal{T}(B)$. Figure 2.6 describes the basic algorithm to compute the marginal and joint image histograms. First, for each of the voxels of the image A, the position in the world coordinates is obtained, taking into account the origin and the spacing of the volume. Then, the point of the moving image is found by applying the inverse transform $\mathcal{T}^{-1}$ to the initial point. If this transformed point does not belong to the image domain, this is not taken into account; otherwise, the intensity value at this point is estimated using a certain interpolation criterion. With this intensity value and the intensity value of the original voxel of the fixed image, the marginal and joint histograms are updated.

```
...
forall voxel in fixedImage do
    inputPoint = fixedImage->TransformIndexToPhysicalPoint(voxel);
    transformedPoint = transform->InverseTransformPoint(inputPoint);
    if( movingImage->IsInside( transformedPoint ) )
        movingValue = interpolator->Evaluate(movingImage, transformedPoint);
        fixedValue = fixedImage->GetValue(voxel);
        histogramX[fixedValue]++;
        histogramY[movingValue]++;
        histogramXY[fixedValue][movingValue]++;
        numberOfPixelsCounted++;
    end if
end forall
...
```

Figure 2.6: Marginal and joint histogram computation algorithm.

Histogram techniques are easily implemented and, probably for that reason, are more commonly used than other non-parametrical techniques, but they have a high computational cost. To overcome this limitation, several approaches has been proposed. The most common is the multiresolution approach, which consists in considering different levels of detail of the images to be registered. Then, a rough estimate of the correct registration is found in relatively little time using downsampled images, which is subsequently refined using images of increasing resolution. The registration process at finer scales uses less number of iterations since the initial solution is closer to the result and, therefore, the overall computational time decreases. Several approaches using this technique are summarized in [132]. Also, some random subsampling strategies have been introduced [10].

2.3.2   PARZEN WINDOW ESTIMATION

Given a set of samples of a random variable, Parzen-windowing [124] estimates the pdf from which the samples are derived by superposing kernel functions placed at each sample value. The general form of the density is

$$\hat{p}(X, a) \equiv \frac{1}{N_a} \sum_{x_a \in a} R(x - x_a),$$

where a is a set of samples, N_a is the cardinality of a, and R is a window function or kernel such that

$$\int_{x \in X} R(x)dx = 1,$$

where X is the domain of function R. This function R is often called the smoothing or window function.

In this way, each observation contributes to the pdf estimate. Unlike parametric estimations, Parzen estimation does not suppose any a priori distribution and only requires the density to be smooth [179]. Intuitively, the Parzen density estimator can be seen as a computation of a windowed average of the sample. The most common window functions are unimodal, symmetric with respect to the origin, and fall off quickly to zero. Among these functions, the Gaussian window is the most popular kernel for Parzen window density estimation, being infinitely differentiable and thereby lending the same property to the Parzen-window pdf estimate. The quality of the approximation is dependent both on the functional form of R and its width. Different window functions can lead to very different density estimates.

In terms of memory, the computation of the Parzen estimation is cheap, since only the set of samples has to be kept in memory. On the contrary, the evaluation of $\hat{p}(x, a)$ is more expensive, requiring a time proportional to the sample size. Moreover, in the joint density function estimation, the computational cost increases with the square of the sample size. Fortunately, the efficiency can be improved if a limited window width is employed [108].

Once the pdf has been estimated, the information measures can be obtained by computing an empirical average based on the observed data. For the continuous entropy (see Equation 1.40), an empirical estimate is given by

$$\hat{h}(X) \quad = \quad \frac{1}{N_b} \sum_{x_b \in b} \log(\hat{p}(x_b, a)),$$

(2.9)

where $\hat{h}(X)$ is the estimation of the continuous entropy of the variable X and b is a second set of samples. Hence, for the estimation of the differential entropy, two set of samples are required: the first one to average the logarithm of the probability of each sample and the second one to estimate the probability of each of the samples of the first set. The joint entropy and mutual information can also be obtained using similar procedures [179, 183]. In addition, the derivatives of these measures, which can help the optimizer to find the best transform in the registration process, can be obtained in a very computationally efficient way [179].

2.3.3 ENTROPIC SPANNING GRAPHS

Another method for a non-parametric estimation of the information measures is based on the entropic spanning graphs [70]. This method does not estimate the probability density function, but directly estimates the Rényi entropy (see Equation 1.57) from the generation of a entropic spanning graph. This type of graphs are constructed on the normalized feature space, where the nodes of the graph correspond to the samples of the variable and the edges represent the euclidian distances between the nodes [70]. Intuitively, the nodes of this graph are closer when the variable has low entropy than when the entropy is high (i.e., the samples of the graph are sparse). It can be seen that there is a relationship between the length of the minimum spanning tree (i.e., the spanning tree with total edge weight less than or equal to the weight of every other spanning tree) and the entropy of the variable. In particular, this relationship is given by

$$\hat{H}_\alpha^R(X) \quad = \quad \frac{1}{1-\alpha} \left(\ln \left(\frac{\mathcal{L}(X)}{n^\alpha} \right) - \ln \beta \right),$$

(2.10)

where $\hat{H}_\alpha^R(X)$ is the estimation of the Rényi entropy with a given α entropic index, $\mathcal{L}(X)$ is the length of the minimum spanning tree for the variable X, n is the number of data samples, and β is a constant bias correction which is independent of the data. Ma et al. [102] proposed to use the minimization of the joint Rényi entropy estimation as a goal measure in the registration process.

2.4 HIGH-DIMENSIONAL INFORMATION MEASURES INCLUDING SPATIAL INFORMATION

The standard information theoretic measures introduced above ignore the spatial information contained in the images. Some works have been focused on overcoming this problem, that is each pair of samples from both datasets are considered for the computation of the information measures independently of its geometrical position or its neighborhood intensity values. Rueckert et al. [145] proposed a second-order MI to incorporate spatial information. This second-order MI is computed from the co-occurrence matrix of an image, which is a 2D histogram with the frequencies of two neighbor grey values in the image. The neighborhood has been defined by the nearest neighbors of each pixel. With this strategy, mutual information is computed considering voxel pairs, and not only single voxel values, capturing in this way a kind of spatial information. The results of this method demonstrate that the robustness of the registration process is increased.

A generalization of this method was presented by Bardera et al. [12], where a the standard information channel $X \to Y$, that considers pairs of samples, is substituted by the channel $X^L \to Y^L$, that considers pairs of data blocks of length L. From this channel, L-dimensional normalized mutual information is defined by

$$NMI(X^L, Y^L) = \frac{I(X^L; Y^L)}{H(X^L, Y^L)} = \frac{H(X^L) + H(Y^L) - H(X^L, Y^L)}{H(X^L, Y^L)}, \tag{2.11}$$

where $H(X^L)$ and $H(X^L, Y^L)$ are given by

$$H(X^L) = - \sum_{x^L \in \mathcal{X}^L} p(x^L) \log p(x^L), \tag{2.12}$$

and

$$H(X^L, Y^L) = - \sum_{x^L \in \mathcal{X}^L, y^L \in \mathcal{Y}^L} p(x^L, y^L) \log p(x^L, y^L). \tag{2.13}$$

$I(X^L; Y^L)$ is the L-dimensional mutual information, and $p(x^L, y^L)$ is the joint probability of the L-dimensional channel. Bardera et al. [12] proposed a random line sampling in order to compute the joint probability matrix. This method is based on uniformly distributed random lines, also called *global lines* [152], which sample the 2D-image or the 3D-volume stochastically in the sense of integral geometry, i.e., invariant to translations and rotations [151]. These lines are generated from the walls of a convex bounding box containing the volume [30]. This can be done taking a random point on the surface of the convex bounding box and a cosinus distributed random direction as it is illustrated in Figure 2.7(a). The sequence of intensity values (L-block X^L) needed to estimate the joint probabilities is captured at evenly spaced positions over the global lines from an initial random offset, that ranges from 0 to the step size (see Figure 2.7(b)). Points chosen on each line provide us with the intensities to calculate the L-block entropies (see Figure 2.7(c)). In this manner, the 3D-neighborhood problem is reduced to 1D, where the concept of neighborhood is well defined.

Other approaches that incorporate more information than only the intensity pairs have been proposed. Pluim et al. [130] include spatial information by combining MI with a term based on the image gradient of the images to be registered. The gradient term seeks to align locations of high gradient magnitude and similar orientations of the gradients at these locations. In this approach

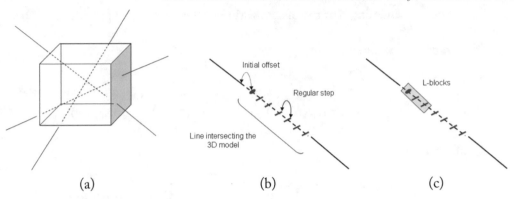

Figure 2.7: (a) Global lines are cast from the walls of the bounding box, (b) intensity values are captured at evenly spaced positions over the global lines from an initial random offset, and (c) neighbor intensity values are taken in L-blocks.

spatial information is incorporated using the gradient values, which measure the difference of a voxel and its neighbors. A more general framework was presented by Butz et al. [25], where MI is computed not from the individual pixel pairs but from the choice of various feature spaces. Sabuncu and Ramadge [147] included spatial information in the MI-based approach by using spatial feature vectors obtained from the images and use a minimum spanning tree algorithm to estimate the conditional entropy in higher dimensions. Russakoff et al. [146] propose an extension of MI which takes into account neighborhood regions of corresponding pixels. They assume that the high-dimensional distribution is approximately normally distributed. Holden et al. [74] used the derivatives of the Gaussian scale space to provide structural information in the form of a feature vector for each voxel. Gan and Chung [58] integrated a maximum distance-gradient-magnitude feature together with the intensity value into a two-element attribute vector and adopt the multidimensional MI as a similarity measure on the vector space.

In all these investigations, the fact of tacking into account the spatial information notably improves the registration results.

2.5   IMAGE REGISTRATION BASED ON f-DIVERGENCES

As it has been introduced in Section 1.9, f-divergences quantify the degree of discrimination between two probability distributions. They are frequently called "distances," although some of them are not strictly metrics.

In Section 2.2.3, mutual information has been presented as a distance between the joint probability $p(X, Y)$ and the distribution that would have been found if the images were completely independent $p(X)p(Y)$. Following this reasoning, some f-divergences have been proposed as image registration measures [133], quantifying the distance between both joint probability functions and assuming that the maximization of these measures provides the correct registration.

The proposed measures are:

- V-information

$$D_V(p(X,Y), p(X)p(Y)) = \sum |p(x,y) - p(x)p(y)|; \qquad (2.14)$$

- I_α-information

$$D_{I\alpha}(p(X,Y), p(X)p(Y)) = \frac{1}{\alpha(\alpha-1)}\left(\sum \frac{p(x,y)^\alpha}{(p(x)p(y))^{\alpha-1}} - 1\right); \qquad (2.15)$$

- M_α-information

$$D_{M\alpha}(p(X,Y), p(X)p(Y)) = \sum |p(x,y)^\alpha - (p(x)p(y))^\alpha|^{\frac{1}{\alpha}}; \qquad (2.16)$$

- χ_α-information

$$D_{\chi\alpha}(p(X,Y), p(X)p(Y)) = \sum \frac{|p(x,y) - p(x)p(y)|^\alpha}{(p(x)p(y))^{\alpha-1}}. \qquad (2.17)$$

It can be shown that for $\alpha \to 1$, $D_{I\alpha}$ converges to the mutual information. Pluim et al. [133] analyzed these measures for different α values in terms of accuracy and robustness. Although some of these measures are less smooth than MI and, therefore, more difficult to optimize, $D_{I\alpha}$ and $D_{M\alpha}$ were shown to occasionally produce more accurate results for the registration of MR and CT images [133].

2.6   SIMILARITY MEASURES BASED ON GENERALIZED ENTROPIES

The most commonly used measure of entropy in the medical image registration context is the Shannon entropy, but other generalized entropies has also been proposed. Wachowiak et al. [180] defined two generalized mutual information measures based on Rényi (see Equation 1.57) and Harvda-Charvát-Tsallis entropy (see Equation 1.58), respectively. The Rényi entropy-based measure was defined as

$$I_\alpha^R(X;Y) = H_\alpha^R(X) + H_\alpha^R(Y) - H_\alpha^R(X,Y), \qquad (2.18)$$

where H_α^R represents the Rényi entropy with a given α parameter. The Tsallis-Havrda-Charvát entropy-based measure was defined as

$$I_\alpha^T(X;Y) = H_\alpha^T(X) + H_\alpha^T(Y) + (1-\alpha)H_\alpha^T(X)H_\alpha^T(Y) - H_\alpha^T(X,Y), \qquad (2.19)$$

where H_α^T represents the Harvda-Charvát-Tsallis entropy with a given α parameter. Results show that, for both measures, α values slightly higher than one (in the range [1.1,2]) obtain the best performance [180].

Bardera et al. [8] proposed the use of these entropies in a different manner: defining a generalized measure based on the Jensen's inequality (see Equation 1.21). From this inequality, the concavity of the entropy functions provides us with the following inequality [23]:

$$J_\alpha^h(\pi_1, \pi_2, \ldots, \pi_n; p_1, p_2, \ldots, p_n) = H_\alpha^h(\sum_{i=1}^n \pi_i \, p_i) - \sum_{i=1}^n \pi_i H_\alpha^h(p_i) \geq 0, \qquad (2.20)$$

where $p_1, p_2, \ldots, p_n$ are a set of probability distributions, $\pi_1, \pi_2, \ldots, \pi_n$ are the priori probabilities or weights, fulfilling $\sum_{i=1}^n \pi_i = 1$, and the superindex h stands for R (Rényi entropy) or T (Tsallis-Havrda-Charvát entropy), depending on the entropy considered. For the Rényi entropy, the parameter α can only take values in the range $0 < \alpha \leq 1$ due to the fact that $H_\alpha^R(X)$ is neither concave nor convex for $\alpha > 1$. The particular case $\pi_i = \frac{1}{n}$ for the Jensen difference applied to the Rényi entropy has been studied by He et al. [66] in medical image registration.

Jensen's divergence coincides with $I(X;Y)$ when $\alpha = 1$, $\{\pi_i\}$ is the marginal probability distribution of X, and $\{p_i\}$ are the rows $\{p(Y|x_i)\}$ of the probability conditional matrix of the channel. Hence, Burbea and Rao [23] defined *generalized mutual information (GMI)* as

$$GMI_\alpha^h(X \to Y) = J_\alpha^h(p(x_1), \ldots, p(x_n); p(Y|x_1), \ldots, p(Y|x_n)). \qquad (2.21)$$

Note that GMI is not symmetric.

The authors in [8] proposed two normalizations of these measures. The first was normalized by the maximum marginal distribution and is given by

$$nGMI_\alpha^h = \max\left\{\frac{GMI_\alpha^h(X \to Y)}{H_\alpha^h(Y)}, \frac{GMI_\alpha^h(Y \to X)}{H_\alpha^h(X)}\right\}. \qquad (2.22)$$

The second normalization of GMI_α^h was obtained by dividing by the joint entropy and it is given by

$$NGMI_\alpha^h = \frac{\max\{GMI_\alpha^h(X \to Y), GMI_\alpha^h(Y \to X)\}}{H_\alpha^h(X, Y)}. \qquad (2.23)$$

The experiments between MR-CT and MR-PET image pairs showed that $nGMI$ and $NGMI$ are more robust than NMI for a determined range of the entropic index. In addition, for a different range of the entropic index, the generalized measures lead to a minimum closer to the optimal registration point than NMI. Depending on the registration modality, the authors suggest to use one range or the other [8].

Moreover, the Jumarie entropy [78] has also been introduced in the image registration field by Rodriguez and Loew [144]. The Jumarie entropy is defined for one-dimensional signals and resembles a normalized version of Shannon entropy, applied not to a probability distribution, but to the differences of neighboring samples. In the medical image registration field, the Jumarie entropy was defined on the gradient magnitude of pixels [144].

2.7   MEASURES BASED ON THE SIMILARITY METRIC

As we have seen in Section 1.11, the *normalized information distance* (NID), also called *the similarity metric* and introduced by Li et al. [95] for measuring similarity between sequences, is based on the non-computable notion of *Kolmogorov complexity* $K(x)$ [96]. NID is a normalized version of the information metric [15] (see Equation 1.64). In essence, the main idea is that two objects are similar if we can significantly compress one given the information in the other. NID has been successfully applied in areas such as genome phylogeny [94], language phylogeny [95], and classification of music pieces [36]. In the image registration context, NID was introduced in [11]. However, the application of NID is limited by its non-computability. To tackle this problem, two different approaches were proposed in the image registration context [11, 13].

The first approach is based on the *normalized compression distance* (NCD) [37], which is a feasible version of the normalized information distance and is computed from the lengths of compressed data files (see Equation 1.65). In this approach, the capability of the compressor to approximate the Kolmogorov complexity will determine the registration accuracy.

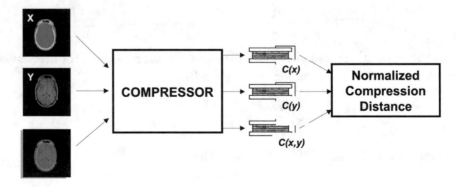

Figure 2.8:   Registration based on compression scheme.

Given two images A and B, the correct registration will be achieved when the normalized compression distance is minimum. In this formula, $C(x)$ (or $C(y)$) represents the size of the compressed image A (or B) and $C(x, y)$ the length of the compressed fused pair $(A, \mathcal{T}(B))$ file (see Figure 2.8). The fusion has been done superimposing the images after applying a certain transform. To compress the images, both image and text compressors can be used [11]. The main drawback of this approach is the feasibility of the real-world compressors to capture the real compressibility of the images.

The second approach proposed in [11] is based on the *normalized entropy rate distance* (*NED*) [46, 79], which substitutes the Kolmogorov complexity in NID (Equation 1.64) by the entropy rate (see Equation 1.29). This is a measure of the degree of compressibility of an image from a Shannon perspective. Thus, the *normalized entropy rate distance* (*NED*) is given by

$$NED(x, y) = \frac{h^{XY} - \min\{h^X, h^Y\}}{\max\{h^X, h^Y\}}, \tag{2.24}$$

where h^X (or h^Y) represents the entropy rate of image A (or $\mathcal{T}(\mathsf{B})$) and h^{XY} the entropy rate of the pair $(\mathsf{A}, \mathcal{T}(\mathsf{B}))$, which is defined as

$$
\begin{aligned}
h^{XY} &= \lim_{L\to\infty} \frac{1}{L} H(X_1, \ldots, X_L, Y_1, \ldots, Y_L) \\
&= \lim_{L\to\infty} H((X_L, Y_L)|(X_{L-1}, Y_{L-1}), \ldots, (X_1, Y_1)), \tag{2.25}
\end{aligned}
$$

represents the maximum compressibility for the two superimposed strings and $H(X_1, \ldots, X_L, Y_1, \ldots, Y_L)$ symbolizes the joint entropy of L symbols of the pairs $(\mathsf{A}, \mathcal{T}(\mathsf{B}))$. With this proposal, neighbor information is used by considering the correspondence between blocks of pixels, instead of the correspondence between individual pixels as in the classical registration methods based on MI [132]. A similar approach was previously proposed by Kaltchenko [79] and Dawy et al. [46] in the bioinformatics field.

For $L = 1$, the entropy rate approximation $H(X_1, Y_1)$ is the standard Shannon joint entropy and, then,

$$NED(x, y)_{L=1} = \frac{H(X, Y) - \min\{H(X), H(Y)\}}{\max\{H(X), H(Y)\}}.$$

Note that the estimation of the entropy rate is more accurate for higher L, due to the high spatial correlation between samples in medical images.

NED can be analyzed from the information channel perspective. The classical registration methods based on MI consider the registration problem as an information channel, $X \to Y$, where the random variables X and Y correspond to both images, and where the mutual information of the channel must be maximized. Here, this information channel is replaced by $X^L \to Y^L$, where $X^L = \{X_1, \ldots, X_L\}$ and $Y^L = \{Y_1, \ldots, Y_L\}$ represent the random variables built from blocks of L pixels. Note that the accuracy of the computation is influenced by the size of the blocks, but the sparsity of the joint histogram and also the computational cost of the process increases with it. Note that this information channel coincides with the one presented in Section 2.4.

Figure 2.9 plots the different values of the *NED* measure considering horizontal translations for three different L values (1, 2, and 3) represented by dash-dotted, solid, and dashed lines, respectively. For comparison purposes, the *NMI* measure is also represented in bold. Due to the high dimensionality of the joint histogram in the $L = 3$ case, the number of bins has been reduced to 8. To preserve the testing conditions, this quantization has been kept in all cases. As we

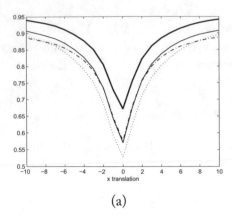

(a)

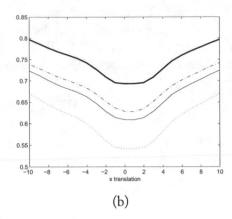

(b)

Figure 2.9:   *NED* measure results for the (a) synthetic MR-T1 and MR-T2 (Figure 2.4) and (b) MR and CT images (Figure 2.1) using different block lengths L (see the description for more details) ([13] © Elsevier, 2010).

expected, entropy rate estimation decreases with L, giving us a more approximated measure of the real entropy rate and, equivalently, the string compressibility. Observe in both plots, the smoothness of the *NED* curves, without local minima, and the accuracy of the registration, achieving their minimum at the correct position for both synthetic MR-T1 and MR-T2 (Figure 2.9(a)) and real MR-CT pairs (Figure 2.9(b)).

2.8   IMAGE FUSION

Multimodal visualization aims at combining the most relevant information from different volumetric data sets into a single visualization that provides as much information as possible [27]. To reach this goal two main steps have to be carried out. First, a registration step is needed to align the input data sets in a common space. Second, a fusion step has to be done to mix the values represented in the same spatial position in order to obtain the final image.

The fusion process is challenging since it requires that the data contained in each pair of registered voxels are reduced into a single visual representation. Different alternatives have been proposed to perform the information merging and generate the fused image. The simplest methods are the checkerboard display, which alternately displays the pixels of the original input data sets, the color channel-based fusion, that assigns the input images to different color channels and then performs the rendering, and the combination of data segmentation (to represent the information of one of the data sets) with color coding (to map the data of the other data set) [27, 56, 185]. These techniques have been proven to be useful in clinical environments, although, in most of the cases, the task of mentally reconstructing the relationship between structures is left to the observer.

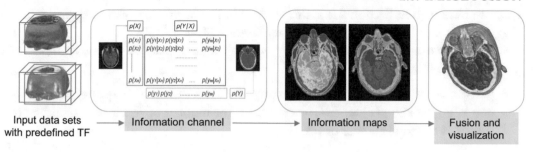

Figure 2.10: Overview of the pipeline for multimodal data fusion proposed by Bramon et al. Adapted from [20] © IEEE, 2012.

Bramon et al. [20] proposed a information-theoretic approach for multimodal visualization that automatically fuses data on the basis of extracting the most relevant information of the input data sets. This approach assumes that both input images are registered and applies a processing pipeline composed of four main steps (see Figure 2.10).

– **Communication channel.** A communication channel is defined between the two registered input data sets. This channel permits to compute the mutual information between these data sets.

– **Specific information.** For each data set, the information associated with each intensity value is computed. This information is also called specific information. Three information measures derived, respectively, from three different decompositions of mutual information are presented. For each source data set, an information map that represents the specific information associated with each intensity value is created.

– **Fusion criteria.** The fused image is obtained from the most informative voxels of the source data sets. That is, for each pair of matched voxels, the most informative one (i.e., with the highest specific information) is selected for the fused data set.

– **Visualization.** Multivolume ray casting is used to render the fused data set. Color and opacities are obtained from the original transfer functions defined for each one of the input data sets.

These four steps are described in the next sections.

2.8.1 COMMUNICATION CHANNEL

The relationship between two multimodal medical images can be represented by a communication channel $X \rightarrow Y$ between the random variables X (input) and Y (output), which represent, respectively, the set of intensity bins $\mathcal{X}$ of the image X and the set of intensity bins $\mathcal{Y}$ of the image

Y. As we have seen in Section 2.2.1, this channel has been previously used to deal with several image processing problems, such as image registration [104, 179].

Given two registered images with N voxels, the three basic components of the channel $X \to Y$ are the following.

- The *input distribution* $p(X)$, which represents the normalized frequency of each intensity bin x, is given by $p(x) = \frac{n(x)}{N}$, where $n(x)$ is the number of voxels corresponding to bin x and N is the total number of voxels.

- The *conditional probability matrix* $p(Y|X)$, which represents the transition probabilities from each bin of the image X to the bins of the image Y, is defined by $p(y|x) = \frac{n(x,y)}{n(x)}$, where $n(x)$ is the number of voxels with intensity x in image X and $n(x, y)$ is the number of voxels with intensity x such that the corresponding voxel in the image Y has intensity y. Conditional probabilities fulfil $\sum_{y \in \mathcal{Y}} p(y|x) = 1, \forall x \in \mathcal{X}$.

- The *output distribution* $p(Y)$, which represents the normalized frequency of each bin y, is given by $p(y) = \sum_{x \in \mathcal{X}} p(x)p(y|x) = \frac{n(y)}{N}$, where $n(y)$ is the number of voxels corresponding to bin y.

2.8.2 SPECIFIC INFORMATION

As we have mentioned in Section 1.3, MI expresses the information gained on a data set by observing the other. In other words, MI quantifies the information contained by the set of intensity values of an image X about the set of intensities of the other image Y. This interpretation can be extended to define, for instance, the information associated with a single intensity value $x \in \mathcal{X}$, that is, the information gained on Y by the observation of an intensity value x. To obtain the specific information associated with an intensity value, MI can be decomposed in different ways [18, 24, 47]. Three forms of decomposing MI have been presented in Section 1.3 to quantify the information associated to a stimulus or a response. In this chapter, these measures are used in the context of an information channel between two images. Although many definitions of specific information are plausible, the three most "natural" decompositions of $I(X; Y)$ are as follows.

- **Surprise** I_1. The definition of *surprise* I_1 can be directly derived from the formula of MI (Equation 1.14), taking the contribution of a single intensity value x to $I(X; Y)$:

$$
\begin{aligned}
I(X; Y) &= \sum_{x \in \mathcal{X}} p(x) \sum_{y \in \mathcal{Y}} p(y|x) \log \frac{p(y|x)}{p(y)} \\
&= \sum_{x \in \mathcal{X}} p(x) I_1(x; Y),
\end{aligned}
\tag{2.26}
$$

where

$$
I_1(x; Y) = \sum_{y \in \mathcal{Y}} p(y|x) \log \frac{p(y|x)}{p(y)}
\tag{2.27}
$$

expresses the surprise about Y from observing x . Observe that surprise $I_1(x;Y)$ is large when $p(y|x)$ is much higher than $p(y)$ for certain values of intensity in Y. The name surprise was used by DeWeese and Meister [47] to emphasize the fact that the observation of x has moved the estimate of y towards values that seemed very unlikely prior to the observation. It can be shown that I_1 is the only positive decomposition of MI [47]. This positivity can be demonstrated by the fact that $I_1(x;Y)$ is the Kullback-Leibler distance (see Equation 1.53) between $p(Y|x)$ and $p(Y)$, which is always positive [42]. Thus, the greater the distance, the greater the surprise.

- **Predictability** I_2. DeWeese and Meister [47] defined the specific information I_2, which we call now *predictability*,[1] from the decomposition of MI obtained from Equation 1.13:

$$\begin{aligned} I(X;Y) &= H(Y) - H(Y|X) \\ &= \sum_{x \in \mathcal{X}} p(x)H(Y) - \sum_{x \in \mathcal{X}} p(x)H(Y|x) \\ &= \sum_{x \in \mathcal{X}} p(x)I_2(x;Y), \end{aligned} \tag{2.28}$$

where

$$\begin{aligned} I_2(x;Y) &= H(Y) - H(Y|x) \\ &= -\sum_{y \in \mathcal{Y}} p(y)\log p(y) + \sum_{y \in \mathcal{Y}} p(y|x)\log p(y|x) \end{aligned} \tag{2.29}$$

expresses the change in uncertainty about Y when x is observed. Note that $I_2(x;Y)$ can take negative values. This means that certain observations x do increase our uncertainty about the state of the variable Y. Intensity values x with high $I_2(x;Y)$ greatly reduce the uncertainty in Y and, thus, they are very significant in the relationship between X and Y. From this interpretation, we can say that I_2 expresses the capacity of prediction of a given intensity value.

- **Entanglement** I_3. Butts [24] defined another measure obtained from the decomposition of MI, called stimulus specific information I_3. This measure, which Bramon et al. call *entanglement*, remembering the mutual correlation between certain quantum systems, is defined by

$$I_3(x;Y) = \sum_{y \in \mathcal{Y}} p(y|x)I_2(y;X). \tag{2.30}$$

A large value of $I_3(x;Y)$ means that the intensity values of Y associated with x are very informative in the sense of $I_2(y;X)$. That is, the most informative input values x are those

[1]Observe that this measure has ben called *specific information* in Section 1.3.

that are related to the most informative outputs y. Observe that $I_1(x;Y)$ and $I_2(x;Y)$ are obtained from both distributions $p(Y)$ and $p(Y|x)$, while $I_3(x;Y)$ is a weighted sum of the measure $I_2(y;X)$, which is obtained from distributions $p(X)$ and $p(X|y)$. This fact will imply a special treatment of the measure I_3 in the next section.

As we have seen, $I_1(x;Y)$, $I_2(x;Y)$, and $I_3(x;Y)$ represent three different ways of quantifying the specific information associated with an intensity value x. The properties of positivity and additivity of these measures have been studied in [18, 24, 47]. A measure is additive when the information obtained about X from two observations, $y \in \mathcal{Y}$ and $z \in \mathcal{Z}$, is equal to that obtained from y plus that obtained from z when y is known. Additivity is a desirable requirement that responds to the intuitive notion that information accumulates additively over a sequence of observations, such that the total obtained in all steps is equal to what we would calculate if we considered all the events as a single observation [47]. While I_1 is always positive and non-additive, I_2 can take negative values but is additive, and I_3 can take negative values and is non-additive. Because of the additivity property, DeWeese and Meister [47] prefered I_2 against I_1 since they consider that additivity is a fundamental property of any information measure. Thus, it can be considered that I_2 is the most natural and intuitive measure of specific information.

Figure 2.11 shows, for the CT and MR-T1 pair (column(a)), the information maps of I_1, I_2, and I_3 (columns(b-c), respectively), tacking as the reference image the CT (row(i)) and the MR-T1 (row(ii)). Observe that the lowest values of I_1 correspond to the parenchyma, and the highest ones are achieved in the extraocular muscles and also in the paranasal sinuses. That is, intensity values from parenchyma and extraocular muscles are associated with "non-surprising" and "surprising" values, respectively. The measure I_2 represents predictability, i.e., how the uncertainty about Y changes when x is observed. Note that while the CT-parenchyma has low values of I_2, the T1-parenchyma has high values. Therefore, the observation of the CT-parenchyma produces a small reduction of the uncertainty over T1 while the observation of T1-parechyma produces a higher reduction of the uncertainty over CT. This is mainly due to the fact that T1 captures information from brain tissues in more detail than CT. Finally, remember that the measure I_3 represents the entanglement, i.e., how the uncertainty about X changes when the intensities y associated with x are observed. As it can be expected from the definition of I_3, we can observe that the regions of interest of each image have low values (e.g., the parenchyma in the T1 image).

2.8.3   FUSION CRITERIA

Bramon et al. proposed two different fusion strategies based on the previously defined measures I_1, I_2, and I_3. First, a *symmetric* method that compares the same specific information measure in both images and, thus, does not prioritize any of the two input images was proposed. Second, an *asymmetric* method that considers two different information measures (I_2 and I_3) of the same data set and, therefore, produces different results depending on the chosen data set (reference data set) was proposed. Next, both methods are described in detail.

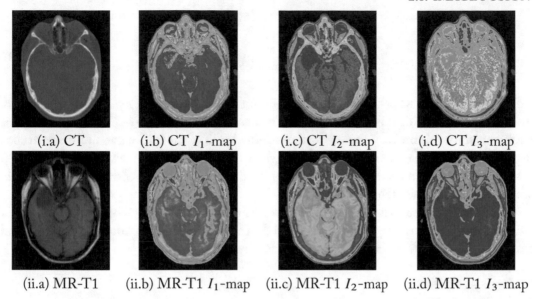

(i.a) CT   (i.b) CT I_1-map   (i.c) CT I_2-map   (i.d) CT I_3-map

(ii.a) MR-T1   (ii.b) MR-T1 I_1-map   (ii.c) MR-T1 I_2-map   (ii.d) MR-T1 I_3-map

Figure 2.11: Information maps obtained for two input data sets corresponding to a (i.a) CT and a (ii.a) MR-T1 images and their corresponding information maps of (b) I_1, (c) I_2, and (d) I_3 in rows (i-ii), respectively ([20] © IEEE, 2012).

Symmetric Method

In these fusion algorithms, a single data set Z, represented by random variable Z, is obtained from the two input data sets X and Y, represented by random variables X and Y, respectively. It is assumed that at each voxel the variable Z takes the value $x \in \mathcal{X}$ or $y \in \mathcal{Y}$ depending on an information optimization criterion. The symmetric fusion strategy is based on the conjecture that, by selecting the most informative intensity values, the most relevant structures of the input data sets will be represented in the fused one. Three symmetric criteria based on the three specific information measures defined in the previous section are now proposed to compute the fused value.

– I_1-*fusion*: For each voxel with input values (x, y), the fused value z is given by

$$z = \begin{cases} x & \text{if } I_1(x; Y) > I_1(y; X) \\ y & \text{otherwise.} \end{cases} \tag{2.31}$$

– I_2-*fusion*: For each voxel with input values (x, y), the fused value z is given by

$$z = \begin{cases} x & \text{if } I_2(x; Y) > I_2(y; X) \\ y & \text{otherwise.} \end{cases} \tag{2.32}$$

 - I_3-*fusion*: For each voxel with input values (x, y), the fused value z is given by

$$z = \begin{cases} x & \text{if } I_3(x; Y) < I_3(y; X) \\ y & \text{otherwise.} \end{cases} \tag{2.33}$$

Observe that, while I_1-fusion and I_2-fusion select the intensity value x when the information associated with x is greater than the one associated with y, I_3-fusion requires opposite treatment since this measure represents the weighted sum of the information of the corresponding intensities in the other data set. That is, when $I_3(x; Y)$ is higher than $I_3(y; X)$, the value y is selected because a high value of $I_3(x; Y)$ means that on average the corresponding values in Y are very informative (in the sense of I_2).

It is important to observe that all the voxels that have the same pair of input intensity values will have the same intensity value in the fused data set. Note also the symmetric role of X and Y that can be permuted without modifying the final result.

Asymmetric Method

In the asymmetric approach, at each voxel, the variable Z takes the value x or y depending on an optimization criterion based on the values of I_2 and I_3 taken from the same data set (reference data set).

As stated above, $I_2(x; Y)$ measures the predictability of the intensity value x over the variable Y, and $I_3(x; Y)$ gathers the predictability of the intensity values of Y associated with intensity x. We introduce the $I_2 I_3$(X)*-fusion* criterion based on the comparison of I_2 and I_3 measures from the reference data set X, where the fused value z is given by

$$z = \begin{cases} x & \text{if } I_2(x; Y) > I_3(x; Y) \\ y & \text{otherwise.} \end{cases} \tag{2.34}$$

Observe that the intensity value x of the reference data set is selected when its predictability is greater than the predictability of the intensity values y associated with x, otherwise the intensity y is selected. This means that the value x is substituted by y when, from the point of view of the reference data set X, the other data set Y provides more information. Note the asymmetric role of X and Y since both measures I_2 and I_3 are taken from X.

2.8.4   VISUALIZATION

Once the fused data set has been generated, it is rendered considering the predefined transfer functions of the input data sets. In this way, the definition of a new transfer function is avoided. This fact notably simplifies the user interaction.

To obtain the final rendering, a standard volume ray casting with an accumulation level mixing scheme is applied. For each ray, at each intersected voxels, the input data set from which the data have to be collected is identified. Then, using tri-linear interpolation, the value of the sample is estimated and both color and opacity values are assigned to this sample from the transfer function of the corresponding input data set.

Finally, Figure 2.12 shows 3D visualizations of the input data sets CT (Figure 2.12(i.a)) and MR-T1 (Figure 2.12(ii.a)) and the corresponding fused data sets using the symmetric fusion methods I_1-fusion (Figure 2.12(i.b)), I_2-fusion (Figure 2.12(i.c)), and I_3-fusion (Figure 2.12(i.d)), and the asymmetric approach $I_2 I_3$-fusion using both input data sets as reference images (Figures 2.12(ii.b-c)). As it can be seen, the visualization of the fused data sets combines the most relevant structures present in each modality.

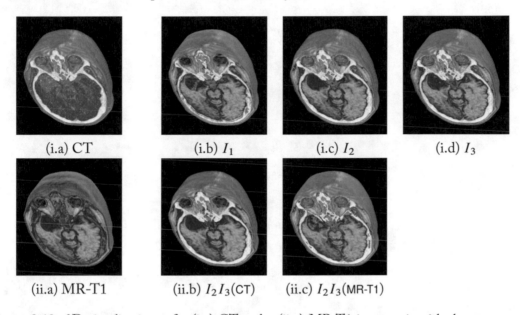

(i.a) CT   (i.b) I_1 (i.c) I_2 (i.d) I_3

(ii.a) MR-T1   (ii.b) $I_2 I_3$(CT)   (ii.c) $I_2 I_3$(MR-T1)

Figure 2.12: 3D visualizations of a (i.a) CT and a (ii.a) MR-T1 image pair with the corresponding 3D fused visualizations applying (i.b-d) the symmetric methods I_1-fusion, I_2-fusion, and I_3-fusion, respectively, and (ii.b-c), the asymmetric methods $I_2 I_3$(CT)-fusion and $I_2 I_3$(MR-T1)-fusion ([20] © IEEE, 2012).

CHAPTER 3

Image Segmentation

Image segmentation is one of the major areas of research in image processing and computer vision. It consists in subdividing an image into its constituent parts and is typically used to identify objects or other relevant information in digital images. Segmentation of non-trivial images is one of the most difficult tasks in image processing, due to the great differences between image types. For instance, natural images have very different features from medical images, from the ones obtained for controlling the quality of manufactured goods, or from scanned documents. Image segmentation has several applications such as analyzing the patient's anatomy from medical images, detecting the layout of an image document, or finding obstacles in a street from a camera installed in a car. In addition, for the same image, different segmentation techniques can be required depending on the regions that the user wants to detect. For instance, if two different anatomical regions have to be segmented in a medical image, then two different segmentation processes will be required.

To overcome the complexity of the segmentation task, several research works have been developed to obtain new segmentation methods as accurate as manual interventions but reducing user interaction as much as possible. Unfortunately, the automation of this process is not easy since the regions to be segmented vary with the images. Consequently, most proposed methods usually assume some *a priori* information that must either be integrated into the system or provided by a human operator. In the image processing literature, we can find a lot of segmentation methods and also very diverse ways of classifying them [51, 55, 60, 126].

In this chapter, the main proposed segmentation algorithms related to information theory are summarized. Section 3.1 presents the maximum entropy thresholding method, which was the first attempt to introduce the information theory to this field, and some extensions of this method. Section 3.2 describes some other thresholding techniques that take into account the spatial distribution of the intensity values in the image to obtain the optimal threshold values. Section 3.3 shows several methods based on evolving curves. Finally, Section 3.4 presents some strategies based on the Information Bottleneck method.

3.1 MAXIMUM ENTROPY THRESHOLDING

Segmentation based on thresholding relies upon the selection of a range of intensity levels, called *threshold values*, for each material class. These intensity ranges are exclusive to a single class, and span the dynamic range of the image. Subsequently, a feature is classified by selecting the class in which the value of the feature falls within the range of feature values of the class. When a single threshold value is considered, the process is called binary thresholding and divides the image into two regions, called foreground (object of interest) and background (non-relevant information). The determination of more than one threshold value is a process called multi-thresholding.

The selection of the threshold generally depends on the visual identification of a peak in the histogram corresponding to a material class, and the selection of a range of intensities around the peak to include only the material class. A possible criterion is to assign the histogram minima as the threshold values. More refined criteria are summarized in [149, 156]. Some information-theoretic measures have successfully been used in thresholding criteria. In this section, we review the most relevant information-theoretic techniques applied to image thresholding. A good survey of these methods is presented by Chang et al. [34].

3.1.1   ENTROPY

In the 1980s, image processing was an emerging area of research and image segmentation was one of the main hot topics. The search of automatic methods to differentiate foreground from background was one of the main focus of research. In this context, Pun [136, 137] proposed the maximum entropy criterion for image thresholding. This was later corrected and improved by Kapur et al. [80].

The maximum entropy thresholding considers the image as a random variable with the normalized image histogram as its probability density function. From here, the optimal threshold value is the one that preserves the maximum information from the original image in terms of Shannon's entropy (Equation 1.2). This optimal threshold maximizes the a priori entropies of the foreground and background classes [80]. From image A, the random variable X takes values in $\{x_1, \ldots, x_n\}$ where n is the number of different intensities of image A. Then, B and F can be defined as the random variables of the background and foreground, respectively. Thresholding methods assume that a thresholding value t separates both regions, in such a way that the variable B takes the values in $\{x_1, \ldots, x_t\}$ and the variable F in $\{x_{t+1}, \ldots, x_n\}$. It is assumed here that the low intensity values of the image correspond to the background and the high ones to the foreground, but this can be easily reversed.

As it has been previously pointed, the probability density function of variable X is obtained from the image histogram. Thus, the pdf of variable B can be computed with the same probabilities of the variable X for the shared states, but divided by a normalizing factor $p(b) = \sum_1^t p_i$ which ensures that the probability distribution sums 1. Similarly, the pdf of the variable F is established from the probabilities of the states $\{x_{t+1}, \ldots, x_n\}$ of the variable X, defining a normalizing factor $p(f) = \sum_{t+1}^n p_i$. Then, the entropies of variables B and F can be, respectively, defined as

$$H(B) \;=\; -\sum_1^t \frac{p_i}{p(b)} \log\left(\frac{p_i}{p(b)}\right) \tag{3.1}$$

(a) (b) $t = 115$

Figure 3.1: (a) The original test image and (b) the resulting binarized image obtained with the maximum entropy thresholding algorithm.

and

$$H(F) \;=\; -\sum_{t+1}^{n} \frac{p_i}{p(f)} \log\left(\frac{p_i}{p(f)}\right), \tag{3.2}$$

where t is the selected threshold. The *maximum entropy thresholding* method looks for threshold t_{op} that maximizes the sum of both entropies:

$$t_{op} \;=\; argmax_t\{H(F) + H(B)\}. \tag{3.3}$$

Experimental results show that this method leads to a thresholded image with a number of black pixels close to the number of white pixels. Figure 3.1 shows a test image from the Berkeley image database [106] and the resulting binarized image using the maximum entropy thresholding algorithm.

The maximum entropy thresholding has been extended to other generalized entropies. First, Sahoo et al. [148] proposed to use Rényi's entropy H_α^R (Equation 1.57) instead of Shannon's one. Thus, this method finds the optimal threshold depending on maximum sum of Rényi's entropies of the foreground and background:

$$t_{op}^{R}(\alpha) \;=\; argmax_t\{H_\alpha^R(F) + H_\alpha^R(B)\}. \tag{3.4}$$

As it can be seen, the optimal threshold depends on the α parameter. Surprisingly, the authors noted that, for $0 < \alpha < 1$, the optimal threshold is always the same and, for $\alpha > 1$, it is also the same, but, in general, different from the one obtained in the range $0 < \alpha < 1$. Moreover, both values were also, in general, different from the one obtained with Shannon's entropy (remember that $lim_{\alpha \to 1} H_\alpha^R(X) = H(X)$).

Portes de Albuquerque et al. [134] proposed the Harvda-Charvát-Tsallis entropy H_α^T (Equation 1.58) in the thresholding criterion. Here, the optimal threshold is defined as

$$t_{op}^{H}(\alpha) \;=\; argmax_t\{H_\alpha^T(F) + H_\alpha^T(B) + (1-\alpha)H_\alpha^T(F)H_\alpha^T(B)\}. \tag{3.5}$$

(a) $H_{0.5}^R$, $t = 146$ (b) $H_{2.0}^R$, $t = 93$

(c) $H_{0.5}^T$, $t = 146$ (d) $H_{2.0}^T$, $t = 93$

Figure 3.2: The results of image 3.1(a) with the generalized entropies method using (a) Rényi's entropy with $\alpha = 0.5$, (b) Rényi's entropy with $\alpha = 2.0$, (c) Harvda-Charvát-Tsallis entropy with $\alpha = 0.5$, and (d) Harvda-Charvát-Tsallis entropy with $\alpha = 2.0$.

The extra term $(1 - \alpha)H_{\alpha}^T(F)H_{\alpha}^T(B)$ is due to the pseudo-additivity entropic rule of the Harvda-Charvát-Tsallis entropy. This rule verifies that, for two independent random variables A and B, the Harvda-Charvát-Tsallis entropy is defined by

$$H_{\alpha}^T(A, B) = H_{\alpha}^T(A) + H_{\alpha}^T(B) + (1 - \alpha)H_{\alpha}^T(A)H_{\alpha}^T(B), \tag{3.6}$$

while the Shannon entropy and the Rényi entropy, which both follow the additivity rule, fulfill that the joint entropy for two independent random variables A and B is given by

$$H(A, B) = H(A) + H(B). \tag{3.7}$$

Figure 3.2 shows the results of applying the generalization with Rényi's and Harvda-Charvát-Tsallis entropies to the maximum entropy thresholding methods to the test image of Figure 3.1(a). For these experiments, $\alpha = 0.5$ and $\alpha = 2.0$ have been used.

3.1.2 RELATIVE ENTROPY

The relative entropy or Kullback-Leibler distance (Equation 1.7) has also been used for image thresholding. This was first proposed by Kittler and Illingworth [83] which assumed that the foreground and background intensities of an image could be modeled by a mixture of two Gaussian distributions. From this model, an estimated random variable $\hat{X}$ can be defined as the mixture of these Gaussian distributions and, hence, the pdf of $\hat{X}$ is given by

$$p(\hat{X}) \quad = \quad \omega p(F) + (1 - \omega)p(B), \tag{3.8}$$

where ω takes values in the range $0 \leq \omega \leq 1$, and $p(F)$ and $p(B)$ are assumed to be the Gaussian distributions

$$p(B) \quad = \quad \frac{1}{\sqrt{2\pi}\sigma_B}e^{-\frac{(x-\mu_B)^2}{2\sigma_B^2}} \qquad (3.9)$$

and

$$p(F) \quad = \quad \frac{1}{\sqrt{2\pi}\sigma_F}e^{-\frac{(x-\mu_F)^2}{2\sigma_F^2}} . \qquad (3.10)$$

For each threshold, the mean and the variance of each class are computed and these define both Gaussian distributions. The value of ω is obtained from the portion of pixels belonging to the foreground and the background for each threshold value. Thus, for each value t a $p(\hat{X})$ is estimated.

But, obviously the real image density $p(X)$ obtained from the image histogram does not exactly coincide with the estimated density $p(\hat{X})$. Kittler and Illingworth proposed the thresholding criterion given by

$$t_{op} \quad = \quad argmin_t\{D_{KL}(p(X); p(\hat{X}))\}, \qquad (3.11)$$

where D_{KL} symbolizes the relative entropy or Kullback-Leibler distance (Equation 1.7). Thus, the best threshold is chosen in such a way that the estimated distribution from the assumption that foreground and background are Gaussian distributed is the most similar to the real image distribution. This method works well if the grey levels of foreground and background are well separated, but this is not always true in many practical applications. Other approaches assume other a priori distributions, such as the Poisson distribution, proposed by Pal and Pal [121].

3.2 THRESHOLDING CONSIDERING SPATIAL INFORMATION

As it has been noted in Section 3.1.1, the maximum entropy thresholding does not take into account the spatial distribution of the samples, but only the image histogram. Consequently, two images with the same histogram will always be segmented with the same optimal threshold, independently on the spatial distribution of the intensity values on these images. In this section, several approaches that consider the spatial relationship between the image intensity values are presented.

3.2.1 GREY-LEVEL CO-OCCURRENCE MATRIX

The first attempt to introduce the spatial information in information-theoretic thresholding methods was based in considering the co-occurrence matrix. A co-occurrence matrix of an image is the joint histogram obtained from the intensity value of a pixel and the value of its neighbors.

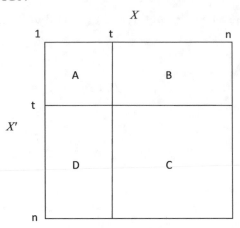

Figure 3.3: The co-occurrence matrix and the four regions (A, B, C, and D) obtained from the threshold value t.

That is, for each image pixel at spatial coordinate (m, n) with its grey level specified by $f(m, n)$, its nearest four neighboring pixels at locations of $(m - 1, n)$, $(m + 1, n)$, $(m, n - 1)$, $(m, n + 1)$ are considered and the corresponding pairs $(f(m, n), f(m - 1, n))$, $(f(m, n), f(m + 1, n))$, $(f(m, n), f(m, n - 1))$ and $(f(m, n), f(m, n + 1))$ are updated during the computation of the joint histogram. This neighborhood definition is referred as the four-adjacency in the work of Gonzalez and Woods [60]. This co-occurrence matrix was first proposed by Haralick [64] for texture analysis purposes. The normalization of the co-occurrence matrix provides the joint probability distribution between the intensity values of a pixel and the value of its neighbors.

The first attempt to introduce this idea for image thresholding is due to Chanda and Majumder [32]. Later, Pal and Pal [120] and Chang et al. [33] proposed a method which subdivides the co-occurrence matrix in four regions depending on the threshold value (see Figure 3.3). Quadrant A denotes the background-to-background transitions, quadrant C the foreground-to-foreground transitions, and B and D the foreground-to-background and the background-to-foreground transitions, respectively. It is assumed that pixels with grey levels above the threshold are assigned to the foreground (corresponding to objects), and those equal to or below the threshold are assigned to the background. Then, the conditional probability of class A can be defined as

$$p^t(x, x'|A) = \frac{p(x, x')}{p^t(A)}, \forall x, x' \in [1, t],$$ (3.12)

where x represents the intensity value of a pixel, x' the intensity of its neighbor, $p(x, x')$ is the transition probability (obtained from the normalization of the co-occurrence matrix), and $p^t(A) = \sum_{x=1}^{t} \sum_{x'=1}^{t} p(x, x')$. Similarly, the conditional probabilities of classes B, C, and D can be defined. The entropy of the conditional probability distributions can be defined as

$$H_t(B, B) = -\sum_{x=1}^{t} \sum_{x'=1}^{t} p^t(x, x'|A) \log p^t(x, x'|A), \qquad (3.13)$$

$$H_t(F, F) = -\sum_{x=t+1}^{n} \sum_{x'=t+1}^{n} p^t(x, x'|C) \log p^t(x, x'|C), \qquad (3.14)$$

$$H_t(F, B) = -\sum_{x=t+1}^{n} \sum_{x'=1}^{t} p^t(x, x'|B) \log p^t(x, x'|B), \qquad (3.15)$$

and

$$H_t(B, F) = -\sum_{x=1}^{t} \sum_{x'=t+1}^{n} p^t(x, x'|D) \log p^t(x, x'|D). \qquad (3.16)$$

Pal and Pal [120] proposed a measure, called *local entropy*, defined as

$$LE(t) = H_t(B, B) + H_t(F, F), \qquad (3.17)$$

and introduced a new segmentation method based on the maximization of this local entropy. This definition can be seen as a second order generalization of the maximum entropy thresholding [80]. Observe that, on the contrary to the standard maximum entropy thresholding, in this case, two images with the same histogram will not, in general, have the same optimal threshold, since the co-occurrence matrix depends on the spatial distribution of the samples.

Pal and Pal [120] defined a new criterion based on the maximization of the transition entropies, that the authors called *joint entropy*. This measure is defined as

$$JE(t) = H_t(B, F) + H_t(F, B). \qquad (3.18)$$

Later, Chang et al. [33] proposed a measure, called *global entropy*, which is defined as

$$\begin{aligned} GE(t) &= LE(t) + JE(t) \\ &= H_t(B, B) + H_t(F, F) + H_t(B, F) + H_t(F, B). \end{aligned} \qquad (3.19)$$

In this case, both local and transition information is taken into account. The authors proposed a thresholding method which finds the threshold that maximizes GE.

Figure 3.4 shows the results of these three methods (LE, JE, and GE maximization) for the test image of Figure 3.1(a).

Another attempt to introduce the spatial information for image thresholding is due to Abu-taleb [4]. In this proposal, the co-occurrence matrix was computed between the intensity of the

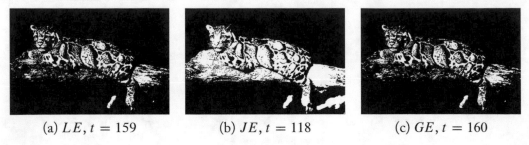

| (a) $LE, t = 159$ | (b) $JE, t = 118$ | (c) $GE, t = 160$ |

Figure 3.4: The results of image 3.1(a) with the co-occurrence matrix method which maximizes (a) LE, (b) JE, and (c) GE.

pixels and mean intensity of the neighboring pixels, which is represented by the variable $\overline{X}$. The normalization of this matrix can be seen as a joint probability $p(X, \overline{X})$. Now, it has to be noted that, on the contrary to the above co-occurrence matrix definition, X and $\overline{X}$ have not the same probability density function. Therefore, in this case, two different threshold values has to be taken into account t and $\overline{t}$ for the variables X and $\overline{X}$, respectively. Thus, similar to the previous approach, a foreground-to-foreground and background-to-background entropies can be defined as

$$H_{t,\overline{t}}(F, F) = -\sum_{x=t+1}^{n} \sum_{\overline{x}=\overline{t}+1}^{n} p_{FF}^{t,\overline{t}}(x,\overline{x}) \log p_{FF}^{t,\overline{t}}(x,\overline{x}) \tag{3.20}$$

and

$$H_{t,\overline{t}}(B, B) = -\sum_{x=1}^{t} \sum_{x'=1}^{\overline{t}} p_{BB}^{t,\overline{t}}(x,\overline{x}) \log p_{BB}^{t,\overline{t}}(x,\overline{x}), \tag{3.21}$$

where

$$p_{FF}^{t,\overline{t}}(x,\overline{x}) = \frac{p(x,\overline{x})}{\sum_{x=t+1}^{n} \sum_{\overline{x}=\overline{t}+1}^{n} p(x,\overline{x})} \tag{3.22}$$

and

$$p_{BB}^{t,\overline{t}}(x,\overline{x}) = \frac{p(x,\overline{x})}{\sum_{x=1}^{t} \sum_{\overline{x}=1}^{\overline{t}} p(x,\overline{x})}. \tag{3.23}$$

Similar to the local entropy LE (see Equation 3.17), the method considers the optimal threshold pair $\{t_{op}, \overline{t}_{op}\}$ which maximizes the sum of both entropies.

3.2.2   MINIMUM SPATIAL ENTROPY THRESHOLDING

The spatial entropy, proposed by Journel and Deutsch [77], measures the sample interdependence depending on the spatial distribution and it can be seen as a measure of the spatial disorder. In particular, this measure is defined as the joint entropy between a sample and another sample at a distance λ. In mathematical terms, the joint probability between these samples is given by

$$p(x, x^\lambda) = \Pr\{x(k), x(k + \lambda)\}, \tag{3.24}$$

where $x(k)$ represents the variable x at the position k and $x(k + \lambda)$ the variable x at the position $k + \lambda$. Then, the spatial entropy can be defined as

$$H^\lambda(X) = -\sum_{x \in \mathcal{X}} \sum_{x^\lambda \in \mathcal{X}} p(x, x^\lambda) log(p(x, x^\lambda)) \geq 0. \tag{3.25}$$

Note that $H^0(X) = H(X, X) = H(X)$ and $H^\infty(X) = 2H(X)$, since, when $\lambda \to \infty$, there is not spatial correlation between x and x^λ. Hence, a normalized measure of the spatial entropy can be written as

$$\overline{H}^\lambda(X) = \frac{H^\lambda(X) - H^0(X)}{H^0(X)} \in [0, 1]. \tag{3.26}$$

Brink [21] proposed the minimization of spatial entropy as a segmentation criterion. The image is treated as a type of Markov random field, where it is assumed that each pixel depends only on the pixels in its immediate neighborhood. Then, the optimal threshold is the one which minimizes the mean spatial entropy for each neighbor of the resulting binarized image. Results indicate a significant improvement over methods that ignore the spatial information.

3.2.3   EXCESS ENTROPY

Another approach that introduces the spatial information for image thresholding was proposed by Bardera et al. [7]. In this work, the excess entropy E (see Equation 1.34) is introduced as a measure of the spatial structure of a 2D or 3D image. Structure here is taken to be a statement which expresses the degree of correlation between the components of a system. Excess entropy, which measures the regularities present in an image, can also be interpreted as the degree of predictability of a pixel given its neighbors. From the concepts introduced in Section 1.5, we analyze how the excess entropy can be computed from Equation 1.34.

In the context of an image, $\mathcal{X}$ represents the set of clusters or bins of the image histogram and x^L is given by a set of L *neighbor* intensity values. In order to compute the excess entropy, two main considerations have to be taken into account:

– The definition of the neighborhood concept for a pixel or voxel. While the concept of neighborhood is unique and unambiguous in 1D, its extension to 2D or 3D introduces ambiguity, since a sequence of L-block neighbor pixels or voxels can be selected in different manners [52].

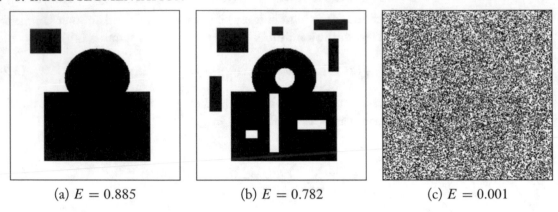

(a) $E = 0.885$ (b) $E = 0.782$ (c) $E = 0.001$

Figure 3.5: Synthetic images and their excess entropy values. The Shannon entropy for all images is the same ([7] © Springer-Verlag, 2009).

- The computation of L-block entropies when $L \to \infty$. In practice, L-block entropies for high L are not computable, since the number of elements of the joint histogram (required to compute joint probabilities $p(x^L)$) is given by N^L, where N is the cardinality of the system. Note that in this case, N is the number of clusters or bins of the segmented image histogram, i.e., the number of colors of the image. Thus, a tradeoff between the accuracy of the measure, given by L, and the number of clusters $|\mathcal{X}|$ is required.

To overcome the neighborhood problem, uniformly distributed random lines, also called *global lines* [152], are used. Global lines have been presented in the registration context (Section 2.4) to stochastically sample a 2D-image or a 3D-volume. The sequence of intensity values (L-block X^L) needed to estimate the joint probabilities is captured at evenly spaced positions over the global lines from an initial random offset, that ranges from 0 to the step size. Points chosen on each line provide us with the intensities to calculate the L-block entropies, required to compute the excess entropy. In this manner, the 3D-neighborhood problem is reduced to 1D, where the concept of neighborhood is well defined. In this implementation, N is taken as an input parameter of the algorithm, while L is determined from N such that the computation of the joint histogram is attainable.

To illustrate the behavior of the excess entropy as a measure of the image structure, three synthetic test images, shown in Figure 3.5, have been generated. The first image (Figure 3.5(a)) represents a synthetic scene with few regions with compact structure. In the second image (Figure 3.5(b)), some additional shapes are added to the original image (Figure 3.5(a)), keeping the same probability for each color. Because of the higher variability of the obtained image, the excess entropy measure decreases, reflecting a lower spatial structure. The last image (Figure 3.5(c)) has been generated by swapping a large number of points of the image of Figure 3.5(a). Each swapping has been done by choosing two random points of the image and interchanging their intensity values. Observe that now the image has not spatial structure (any shape can be observed) and, therefore, the excess entropy is close to 0. It is important to remark that the values of the Shannon entropy of all the images of Figure 3.5 are the same, since the probabilities of each color have remained unaltered.

From the assumption that an image is structured in regions, this method is based on the conjecture that the optimal thresholding should provide us with the maximum structure [7]. Consequently, the selection of thresholds was formulated as a histogram quantization problem using the maximization of excess entropy. That is, the optimal histogram thresholds should correspond to the maximum excess entropy of the resulting image.

3.3 EVOLVING CURVES

The methods based on evolving curves deform an initial curve to fit the desired segmented object following certain energy function which can depend on intensity, texture, or shape. Different approaches have been proposed such as snakes [81], balloons [39], or geodesic active contours [29]. One of the main drawbacks of these methods are the problems related with the topological changes of the curve. In order to solve these problems, the level sets method [118] was proposed. The main difference of this method compared to the previous ones is that, instead of wrapping an initial curve, at each iteration, it deforms a higher-dimensional function depending on the a priori information of the model. The zero-level of this higher-dimensional function is considered as the segmentation curve. On the contrary to the thresholding techniques, with this kind of methods, a pixel is segmented depending on its position on the image instead of its intensity value.

In order to define the energy function required to the curve evolution, some information-theoretic approaches have been proposed, such the ones by Herbulot et al. [68, 69] and by Kim et al. [82].

In the work proposed by Herbulot et al. [68], the concept of entropy is used to characterize the uniformity of the intensity inside the curve. Thus, they define the functional to be minimized using the expression

$$E(\Omega) = \int_{\Omega} -p(I(x), \Omega) \ln p(I(x), \Omega) dx, \qquad (3.27)$$

where Ω is the region inside the curve and $p(I(x), \Omega)$ is the probability of the intensity $I(x)$ at a given point x inside the region Ω. As it can be seen, the continuous space is used. In order to estimate the pdf, the non-parametric approach of the Parzen window is used. The method establishes a region competition defining the same functional for the outside region. The generalization to color images and motion vectors using joint entropies instead of the marginal one is also presented in [69]. Divergence measures, such as the Kullback-Leibler divergence, has also been used to define this energy functional [91]. Figure 3.6 shows a result of this approach. Figures 3.6(a) and 3.6(b) show, respectively, the original curve and the associated histograms for the

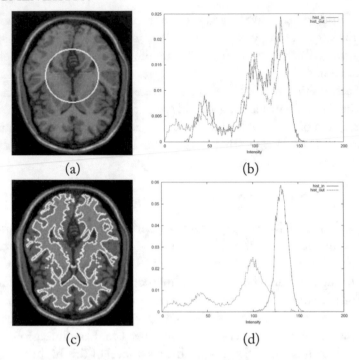

Figure 3.6: (a) The original curve and (b) the associated histograms for the interior and exterior regions, and (c) the final curve and (d) final histograms after applying the Kullback-Liebler energy functional. Courtesy of Dr. Jehan-Besson. From [91] ([91] © Springer-Verlag, 2009).

interior and exterior regions. Figures 3.6(c) and 3.6(d) represent, respectively, the final curve and histograms after applying the Kullback-Liebler energy functional.

An extension of this method was proposed by Duay et al. [49]. In their approach, the prior knowledge obtained from an atlas was introduced in the computation of the region entropy. The method optimizes an energy function designed to be minimal when the entropy of the inside and outside regions of the evolving active contour were close to those of a reference image or atlas.

In the work by Kim et al. [82], an information channel between the regions and the image intensities is defined. The probabilities of the regions are given for their relative area with respect to the total image area. Then, the mutual information (MI) between regions and intensities is used as the curve energy functional. The Parzen window approach is also used to estimate the intensity density functions as well as the gradient flow, required for the curve evolution estimation. A generalization to multiple curves is also presented. A similar functional is obtained from a different point of view by Herbulot et al. [69].

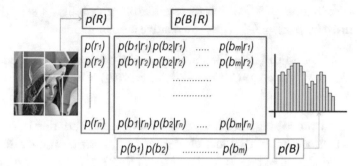

Figure 3.7: The information channel between the regions of the images (R) and the intensity bins (B) for the split-and-merge algorithm. The reverse of this channel used in the histogram clustering algorithm ([14] © IEEE, 2009).

3.4   INFORMATION BOTTLENECK METHOD FOR IMAGE SEGMENTATION

The information bottleneck method is a clustering algorithm based on the maximization of mutual information between two random variables that constitute the input and output of an information channel (see Section 1.8). As it has been seen in the previous section, an information channel can be defined between regions of the image and their intensity values. In [14], the definition of multiple information channels (among them this latter) and the information bottleneck method are used in order to define multiple image segmentation schemes. Next, these approaches are summarized.

3.4.1   SPLIT-AND-MERGE ALGORITHM

The first approach that introduces the information bottleneck method for image segmentation was proposed by Rigau et al. [139]. In this approach, an information channel $R \rightarrow B$ between the random variables R (input) and B (output), which represent, respectively, the set of regions $\mathcal{R}$ of an image and the set of intensity bins $\mathcal{B}$ (see Figure 3.7) is presented. This information channel is defined by a conditional probability matrix $p(B|R)$ which each row of this matrix is given by the normalized intensity histogram of the corresponding row. Thus, this matrix expresses how the pixels corresponding to each region of the image are distributed into the histogram bins.

   Given an image with N pixels, N_r regions, and N_b intensity bins, the three basic elements of the channel $R \rightarrow B$ are as follows.

– The conditional probability matrix $p(B|R)$, which represents the transition probabilities from each region of the image to the bins of the histogram, is defined by $p(b|r) = \frac{n(r,b)}{n(r)}$, where $n(r)$ is the number of pixels of region r and $n(r,b)$ is the number of pixels of region r corresponding to bin b. Conditional probabilities fulfil $\sum_{b \in \mathcal{B}} p(b|r) = 1, \forall r \in \mathcal{R}$.

- The input distribution $p(R)$, which represents the probability of selecting each image region, is defined by $p(r) = \frac{n(r)}{N}$ (i.e., the relative area of region r).

- The output distribution $p(B)$, which represents the normalized frequency of each bin b, is given by $p(b) = \sum_{r \in \mathcal{R}} p(r) p(b|r) = \frac{n(b)}{N}$, where $n(b)$ is the number of pixels corresponding to bin b.

From the data processing inequality (Equation 1.27) and the information bottleneck method (Section 1.8), we know that any clustering or quantization over R or B, respectively represented by $\widehat{R}$ and $\widehat{B}$, will reduce $I(R; B)$. Thus, $I(R; B) \geq I(R; \widehat{B})$ and $I(R; B) \geq I(\widehat{R}, B)$.

Splitting

The splitting phase of the algorithm is a greedy top-down procedure which partitions an image in quasi-homogeneous regions [139]. The partitioning strategy takes the full image as the unique initial partition and progressively subdivides it (e.g., with vertical or horizontal lines) chosen according to the maximum MI gain for each partitioning step. In those experiments, binary space partitioning (BSP) and quad-tree strategies will be used. Note that similar algorithms have been introduced in the context of pattern recognition [155], learning [88], and DNA segmentation [17]. In particular, this algorithm has been used to analyze the complexity of art paintings [141] and, specifically, Van Gogh's work [142] (see Chapter 5 for more details).

The partitioning process is represented over the channel $\widetilde{R} \to B$, where $\widetilde{R}$ denotes that R is the variable to be partitioned. Note that this channel varies at each partition step because the number of regions is increased and, consequently, the marginal probabilities of $\widetilde{R}$ and the conditional probabilities of $\widetilde{R}$ known B also change. For a BSP strategy, the gain of MI due to the partition of a region $\tilde{r}$ in two neighbor regions r_1 and r_2, such that

$$p(\tilde{r}) = p(r_1) + p(r_2) \tag{3.28}$$

and

$$p(b|\tilde{r}) = \frac{p(r_1) p(b|r_1) + p(r_2) p(b|r_2)}{p(\tilde{r})}, \tag{3.29}$$

is given by

$$
\begin{aligned}
\delta I_{\tilde{r}} &= I(R; B) - I(\widetilde{R}; B) \\
&= p(\tilde{r}) JS(\pi_1, \pi_2; p(B|r_1), p(B|r_2)),
\end{aligned} \tag{3.30}
$$

where $\pi_1 = \frac{p(r_1)}{p(\tilde{r})}$ and $\pi_2 = \frac{p(r_2)}{p(\tilde{r})}$. The JS-divergence $JS(\pi_i, \pi_j; p(B|r_1), p(B|r_2))$ (see Equation 1.25) between two regions can be interpreted as a measure of *dissimilarity* between them respect to the intensity values. That is, when a region is partitioned, the gain of MI is equal to the degree of dissimilarity between the resulting regions times the size of the region. In this splitting algorithm, the optimal partition is determined by the maximum MI gain $\delta I_{\tilde{r}}$.

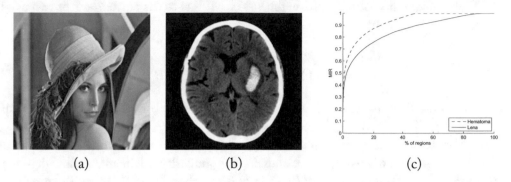

Figure 3.8: Test images: (a) Lena and (b) Hematoma. The two plots in (c) show the mutual information ratio (MIR_r) with respect to the number of regions for (a) and (b) ([14] © IEEE, 2009).

The BSP partitioning algorithm can be represented by an evolving binary tree where each leaf corresponds to a terminal region of the image [155]. At each partitioning step, the tree gains information from the original image such that each internal node k contains the information I_k gained with its corresponding splitting. At a given moment, $I(R; B)$ can be obtained adding up the information available at the internal nodes of the tree weighted by $p(k)$, where $p(k) = \frac{n(k)}{N}$ is the relative area of the region associated with node k and $n(k)$ is the number of pixels of this region. Thus, the MI of the channel is given by

$$I(R; B) = \sum_{k=1}^{T} p(k)I_k, \tag{3.31}$$

where T is the number of internal nodes. It is important to stress that the best partition can be decided locally. That is, the information gained I_k in a given node k is independent of the level of partitioning of the other regions of the image.

From Equation 1.9, the partitioning procedure can also be visualized as $H(B) = I(R; B) + H(B|R)$, where $H(B)$ is the histogram entropy and $I(B; R)$ and $H(B|R)$ represent, respectively, the successive values of MI and conditional entropy obtained after the successive partitions. The progressive acquisition of information increases $I(R; B)$ and decreases $H(B|R)$. This reduction of conditional entropy is due to the progressive homogenization of the resulting regions. Observe that the maximum MI that can be achieved is the histogram entropy $H(B)$, that remains constant along the process. The partitioning algorithm can be stopped using a ratio $MIR_r = \frac{I(R;B)}{H(B)}$ of mutual information gain or a predefined number of regions N_r.

Merging

From the agglomerative information bottleneck method [160] applied to the channel $R \rightarrow B$, we know that any clustering over R will not increase $I(R; B)$. Analogous to the MI gain (Equation 3.30) obtained in the splitting phase, the loss of MI due to the clustering $\hat{r}$ of two neighbor regions r_1 and r_2 is given by

$$
\begin{aligned}
\delta I_{\hat{r}} &= I(R; B) - I(\widehat{R}; B) \\
&= p(\hat{r})JS\left(\pi_1, \pi_2; p(B|r_1), p(B|r_2)\right),
\end{aligned}
\tag{3.32}
$$

where $\widehat{X}$ denotes that the variable X has been clustered, $p(\hat{r}) = p(r_1) + p(r_2)$, $p(b|\hat{r}) = \frac{p(r_1)p(b|r_1)+p(r_2)p(b|r_2)}{p(\hat{r})}$, $\pi_1 = \frac{p(r_1)}{p(\hat{r})}$, and $\pi_2 = \frac{p(r_2)}{p(\hat{r})}$.

As it has seen in the splitting phase, the JS-divergence between two regions can be interpreted as a measure of *dissimilarity* between them. The similarity will be maximum when the two regions have the same histogram: if $p(B|r_1) = p(B|r_2)$, then $\delta I_{\hat{r}} = 0$. Thus, if two regions are very similar (i.e., the JS-divergence between them is small) the channel could be simplified by substituting these two regions by their merging, without a significant loss of information. This is the principle that leads to the following merging algorithm.

From a given image partitioning, the algorithm merges successively the pairs (r_1, r_2) of neighbor regions such that $\delta I_{\hat{r}}$ is minimum. Thus, the number of regions decreases progressively together with the MI of the channel. Similarly to the splitting algorithm, the stopping criterion can be determined by the ratio $MIR_r = \frac{I(R;B)}{H(B)}$ or a predefined number of regions. Note that any initial image partition could be used as input of this algorithm, although the most natural strategy consists in using the result of the algorithm presented in the previous section as input of the merging step. For instance, Calderero and Marqués [28] proposed similar information theory-based measures as a merging strategy of a given image partition.

Note that the clustering $\widehat{R}$ of all regions would give $I(B; \widehat{R}) = 0$. From (1.9), during the merging process $H(B) = I(B; \widehat{R}) + H(B|\widehat{R})$, where $I(B; \widehat{R})$ and $H(B|\widehat{R})$ represent, respectively, the successive values of MI and conditional entropy obtained after the successive mergings. Remember that $H(B)$ remains constant. Note also that $H(B|\widehat{R})$ is the average entropy of the regions, given by

$$
\begin{aligned}
H(B|\widehat{R}) &= -\sum_{r \in \mathcal{R}} p(r) \sum_{b \in \mathcal{B}} p(b|r) \log p(b|r) \\
&= -\sum_{r \in \mathcal{R}} p(r) H(B|r),
\end{aligned}
\tag{3.33}
$$

where $H(B|r)$ is the entropy of the normalized histogram of region r. If two regions are clustered:

$$
\delta I_{\hat{r}} = I(R; B) - I(\widehat{R}; B) = H(B|\widehat{R}) - H(B|R).
\tag{3.34}
$$

Thus, $H(B|\widehat{R})$ never decreases at any iteration due to the mixing of the histogram regions.

In Figure 3.9, the results of merging the regions of the images of Figures 3.8(a) and 3.8(b) obtained from the splitting phase with a $MIR_r = 0.8$ in the BSP partition are shown. For both images, the results with 6 and 10 different regions are shown.

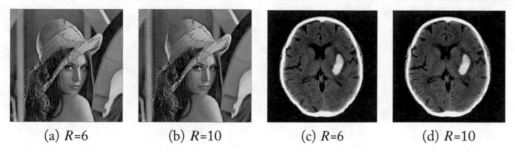

(a) *R*=6   (b) *R*=10   (c) *R*=6   (d) *R*=10

Figure 3.9: Segmentation results of the split-and-merge algorithm for the Lena image (Figure 3.8(a)) and Hematoma image (Figure 3.8(a)), where *R* represents the final number of regions of each image ([14] © IEEE, 2009).

3.4.2   HISTOGRAM CLUSTERING

In this section, a greedy histogram clustering algorithm which takes as input a partitioned image and obtains a histogram clustering based on the minimization of the loss of MI is presented. That is, the bins of the histogram are grouped so that the MI is maximally preserved. From the perspective of the information bottleneck method, the binning process is controlled by a given partition of the image. This histogram clustering algorithm has been previously presented in [139]. Since in this approach only contiguous bins are considered, the method can be seen as a multi-thresholding technique.

This clustering algorithm is based on the channel $B \to R$, which is a result of inverting the channel of the previous section. This channel is defined by a conditional probability matrix $p(R|B)$ which expresses how the pixels corresponding to each histogram bin are distributed into the regions of the image. Bayes' theorem, expressed by $p(b)p(r|b) = p(r)p(b|r)$, establishes the relationship between the conditional probabilities of both channels $B \to R$ and $R \to B$.

The basic idea underlying the histogram clustering algorithm is to capture the maximum information of the image with the minimum number of histogram bins. Analogous to the merging algorithm of the previous section, the loss of MI due to the clustering $\widehat{b}$ of two neighbor bins b_1 and b_2 is given by

$$\begin{aligned} \delta I_{\widehat{b}} &= I(B;R) - I(\widehat{B};R) \\ &= p(\widehat{b}) JS(\pi_1, \pi_2; p(R|b_1), p(R|b_2)), \end{aligned} \tag{3.35}$$

where $p(\widehat{b}) = p(b_1) + p(b_2)$, $p(r|\widehat{b}) = \frac{p(b_1)p(r|b_1)+p(b_2)p(r|b_2)}{p(\widehat{b})}$, $\pi_1 = \frac{p(b_1)}{p(\widehat{b})}$, and $\pi_2 = \frac{p(b_2)}{p(\widehat{b})}$. Thus, when two neighbor bins b_1 and b_2 are equally distributed in the regions of the image ($p(R|b_1) = p(R|b_2)$), their clustering results in $\delta I_{\widehat{b}} = 0$. In general, if two bins are very *similar* ($\delta I_{\widehat{b}} \approx 0$), the channel can be simplified by substituting these two bins by their clustering, without a significant loss of information. The algorithm proceeds by merging two neighbor bins so

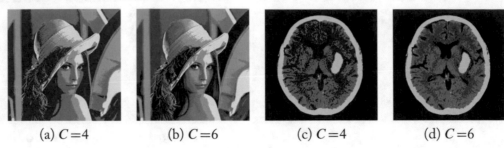

(a) $C=4$ (b) $C=6$ (c) $C=4$ (d) $C=6$

Figure 3.10: Segmentation results of the histogram clustering algorithm for the Lena image (see Figure 3.8(a)) and Hematoma image (see Figure 3.8(b)), where C represents the final number of intensity bins of each image ([14] © IEEE, 2009).

that the loss of MI is minimum. The stopping criterion is given by the ratio $MIR_b = \frac{I(\widehat{B};R)}{I(B;R)}$ or a predefined number of bins N_b.

Note that, during the clustering process $H(R) = H(R|\widehat{B}) + I(\widehat{B};R)$, where $H(R)$ is the entropy of $p(R)$, and $H(R|\widehat{B})$ and $I(\widehat{B};R)$ represent, respectively, the successive values of conditional entropy and MI obtained after the successive clusterings. Observe also that $H(R|\widehat{B})$ is the average entropy of the bins (i.e., a measure of the degree of dispersion of the bins in the set of regions) and increases (or remains constant) at each iteration.

In Figure 3.10, the segmented images obtained from the partitions achieved with the split-and-merge algorithm with $MIR_r = 0.8$ as stopping criterion of the splitting process and 100 regions for the merging one are shown. For each image, the results obtained using four and six clusters are shown. For instance, observe how the internal structures of the brain are approximately preserved using only six clusters.

3.4.3   REGISTRATION-BASED SEGMENTATION

In this section, two histogram clustering algorithms based on the channel established between two registered images A and B are introduced. The main idea behind this algorithms is that the segmentation of image A is obtained by extracting the structures that are most relevant for image B. In this case, any previous segmentation is required. This approach can also be applied to any image with two kind of information per pixel, for instance, an infrared image and a visible light one. These histogram clustering algorithms have been introduced in [9].

One-sided Clustering Algorithm
This algorithm consist in a greedy hierarchical clustering algorithm that clusters the histogram bins of image A by minimizing the loss of MI between A and B. First of all, in a preprocessing step, images A and B have to be registered, establishing an information channel $X \rightarrow Y$, where X and Y denote, respectively, the histograms of A and B. From the data processing inequality

(Equation 1.27) and the information bottleneck method (see Section 1.8), we know that any clustering over X (for instance, merging neighbor histogram bins x_1 and x_2), denoted by $\widehat{X}$, will reduce $I(X; Y)$.

At the initial stage of the algorithm, only one intensity value is assigned to each histogram bin of X. Then, the algorithm proceeds greedily by merging two neighbor clusters so that the loss of MI is minimum. This procedure merges the two most similar clusters from the perspective of B. Note the constraint that only neighbor bins can be merged. The cardinality $|\widehat{X}|$ goes from $|X|$ to 1 in the extreme case.

The efficiency of this algorithm can be greatly improved if the reduction of MI due to the merging of bins x_1 and x_2 is computed by

$$\delta I_{\hat{x}} = p(\hat{x}) JS(\pi_1, \pi_2; p(Y|x_1), p(Y|x_2)), \tag{3.36}$$

where $p(\hat{x}) = p(x_1) + p(x_2)$, $\pi_i = \frac{p(x_1)}{p(\hat{x})}$, $\pi_2 = \frac{p(x_2)}{p(\hat{x})}$, and $p(Y|x_1)$ and $p(Y|x_2)$ denote, respectively, the corresponding rows of the conditional probability matrix of the information channel [160]. The evaluation of $\delta I_{\hat{x}}$ for each pair of clusters is done in $O(|Y|)$ operations and, at each iteration of the algorithm, it is only necessary to compute the $\delta I_{\hat{x}}$ of the new cluster with its two corresponding neighbors. All the other precomputed $\delta I_{\hat{x}}$ values remain unchanged [160].

Similar to the algorithms of Sections 3.4.1 and 3.4.2, clustering can be stopped using several criteria: a fixed number of clusters, a given ratio $MIR_b = \frac{I(\widehat{X};Y)}{I(X;Y)}$, or a variation $\delta I_{\hat{x}}$ greater than a given ϵ. The MIR_b ratio is considered as a quality measure of the clustering.

In Figure 3.11, the rows (i) and (iv) show the results of the one-sided clustering algorithm applied to MR-T1 and MR-T2 registered images from the Brainweb database using 4, 5, and 6 clusters.

Co-clustering Algorithm

Let us now consider a simultaneous clustering of images A and B. Unlike the algorithm presented by Dhillon [48] for word-document clustering, which alternatively clusters the variables $\widehat{X}$ and $\widehat{Y}$, the algorithm chooses at each step the best merging of one of the two images (i.e., the one that entails a minimum reduction of MI). The similarity between the two images is being symmetrically exploited. Thus, each clustering step benefits from the progressive simplification of the images. One of the main advantages of this algorithm is the great reduction of sparseness and noise of the joint probability matrix.

From the data processing inequality (Equation 1.27), $I(\widehat{X}; \widehat{Y})$ is a decreasing function with respect to the reduction of the total number of clusters $|\widehat{X}| + |\widehat{Y}|$. Thus, $I(\widehat{X}; \widehat{Y}) \leq I(X; Y)$. Like the one-sided algorithm, the stopping criterion can be given by a predefined number of bins, a given ratio $MIR = I(\widehat{X}; \widehat{Y})/I(X; Y)$ or a variation $\delta I_{\hat{x}}$ (or $\delta I_{\hat{y}}$) greater than a given ϵ. Similar to the above one-sided algorithm, the reduction of MI can be computed from the JS-divergence (see Equation 3.36). But in the co-clustering algorithm, for each clustering of $\widehat{X}$ (or $\widehat{Y}$), it is necessary to recompute all the $\delta I_{\widehat{y}}$ (or $\delta I_{\widehat{x}}$). The stopping criterion of the co-clustering algorithm is given by the total number of clusters.

In Figure 3.11, the results of the co-clustering algorithm applied to MR-T1 and MR-T2 registered images from the Brainweb database using 4, 5, and 6 clusters are presented in rows (ii) and (iii).

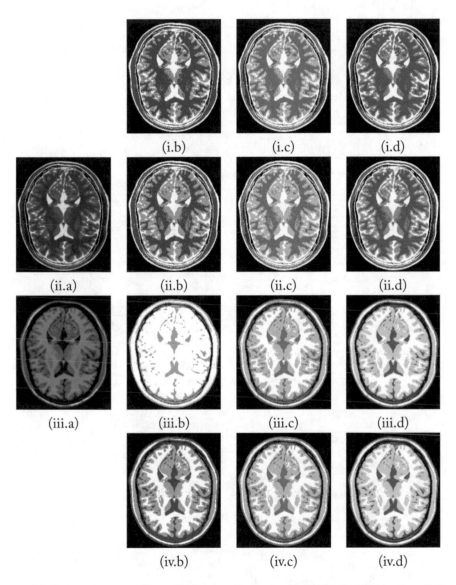

Figure 3.11: (a) Original images from the Brainweb database with 3% of noise. (b,c,d) Images segmented using 4, 5, and 6 bins, respectively. (i,iv) Images obtained with the one-sided algorithm. (ii,iii) Images obtained with the co-clustering algorithm ([14] © IEEE, 2009).

CHAPTER 4

Video Key Frame Selection

Digital videos have become more and more popular in our current society, thanks to the development of video capture devices. A lot of real applications require having a good and quick understanding of general video sequences. But this task is not trivial, due to the fact that understanding videos means essentially dealing with a huge amount of data. Video key frames, which are a set of images representing the main content of the original video data, can be utilized for this purpose. As a result, key frame extraction is a fast and sound way for summarizing a video sequence. Notice that the segmentation of a video sequence into its constituent shots, namely the shot boundary detection, is often a fundamental step for key frame selection. Here a video shot can be defined as a sequence of frames captured by one camera in a single continuous action in time and space [99].

In this chapter, we present several efficient information theoretic techniques for video key frame selection. These approaches are mainly based on mutual information (MI) and its corresponding extensions to the Tsallis mutual information (TMI), and the Jensen-Shannon divergence (JS) and its extensions to the Jensen-Rényi divergence (JR) and the Jensen-Tsallis divergence (JT). These measures are investigated to estimate the frame-by-frame distance or similarity between consecutive video images, for recognizing shot or subshot boundaries and for choosing key frames.

4.1 RELATED WORK AND FIRST IT-BASED APPROACHES

One of the most popular approaches to key frame selection is performed through video shot detection. As mentioned above, a shot is often defined as a sequence of frames that was continuously captured from a single camera at a time, and it can contain pans, tilts, or zooms. Usually, a shot is a group of frames that have consistent visual characteristics such as color, texture, and motion. A video sequence normally contains a large number of shots, which are connected with each other through different video editing methods. A shot boundary is the gap between two shots. The main difficulties of automatic shot boundary detection are related to the object motion and illumination variations, which can be easily confused with a shot boundary. The transitions between shots can be classified into two basic groups: hard or abrupt cuts and gradual transitions. A hard cut is an abrupt shot change that occurs between two continuous frames, i.e., when the last frame in one shot is followed by the first frame in the next shot. Gradual transitions occur over multiple frames and the most typical are fade-in, fade-out, dissolve, and wipe, but many other types of gradual transition are possible [63, 98, 189].

As a matter of fact, many key frame selection algorithms have been introduced in the literature [114, 170]. Among them, the classic information theory metrics, including Shannon entropy, mutual information (MI), Jeffrey (or symmetric K-L) divergence and Kullback-Leibler divergence, have demonstrated their good capability of calculating the similarity or, inversely, the difference between video frames for key frame selection in any kind of video sequence [26, 31, 75, 109, 117]. Next, we review these previous methods that specifically use the information theoretic measures for key frame extraction.

Cernekova et al. [31] presented a shot detection method based on mutual information and joint entropy between frames. Abrupt shot cuts are detected using MI between frames while fade-outs and fade-ins are detected using joint entropy between frames. This approach achieves very good results using an adaptive thresholding based on the local mutual information values in a temporal window. In [31], MI is used as a similarity measure between neighboring images, and shot cuts corresponding to low MI values are detected. Cernekova et al. defined the similarity between two consecutive frames i and j as

$$I^{RGB}(i;j) = I^R(i;j) + I^G(i;j) + I^B(i;j),$$ (4.1)

where the superindices R, G, and B stand for the red, green, and blue color components, respectively, and $I^c(i;j)$ is the mutual information (Equation 1.9) between frames i and j for a given color component c. The marginal probabilities $p(x)$ and $p(y)$ used in the computation of mutual information are given by the normalized histograms of the corresponding color component of frames i and j, respectively, and the joint probability $p(x, y)$ is given by the probability of finding the value x in the frame i and the value y in the frame j at the same pixel location. Cernekova et al. [31] also proposed a ratio between the mutual information $I^{RGB}(i;j)$ and its average value in the neighborhood of pair (i, j) in order to capture relative variations of $I^{RGB}(i;j)$ with respect to the surrounding frames. This ratio is defined as

$$IR^{RGB}(i;i+1) = \frac{I^{RGB}(i;i+1)}{\frac{1}{2r}\sum_{j=i-r,j\neq i}^{i+r} I^{RGB}(j;j+1)},$$ (4.2)

where r is the radius of the window centered in the transition between frame i and $i + 1$ and given by the frames $\{i - r, \ldots, i - 1, i + 1, \ldots, i + r\}$ (see Figure 4.1). In fact, the ratio defined here expresses the video content changes between consecutive frames. Then key frames are extracted from each shot according to how large the ratio is. In the case of smaller content changes within a shot, the first or a middle frame is selected as the key frame. In the case of bigger content changes within a shot, a split-merge procedure is carried out to choose several key frames.

Note that Butz and Thiran [26] also presented an approach that uses mutual information in the gray-scale space to measure changes between subsequent frames in image sequences. In order to compensate camera panning and zooming, Butz and Thiran used affine image registration, and this is the main difference between their method and that by Cernekova et al. [31]. However, doing image registration makes their approach computationally expensive.

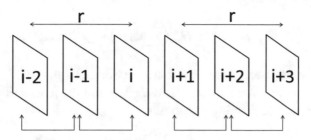

Figure 4.1: Set of frames with radius r centered in the transition between frame i and $i + 1$ ([178] © Springer, 2013).

Janvier et al. [75] reduced the key frame selection to the obtention of a solution to a cost function, established by using the Jeffrey divergence and a minimum message length criterion in information theory.

Omidyeganeh et al. [117] employed the frame-by-frame distance evaluated from the Kullback-Leibler divergence on generalized Gaussian density parameters of wavelet coefficients of video images for separating shots and for obtaining the key frames.

We should note that although most methods use the RGB color space or the luminance value for analyzing the video sequence, some works have studied other color spaces. For instance, Gargi et al. [59] investigated the efficacy of some methods for cut detection and the effect of color space representation on the performance of histogram-based shot detection. Zhang et al. [191] also used the HSV histogram differences of two consecutive frames as a feature for evaluating the color information. We will discuss in Section 4.3 the use of these additional color spaces in the context of information-theoretic measures with results presented in Section 4.4.2.

4.2 KEY FRAME SELECTION BASED ON JENSEN-SHANNON DIVERGENCE AND JENSEN-RÉNYI DIVERGENCE

Xu et al. analyzed the behaviour of Jensen-Shannon divergence and its extension to Jensen-Rényi divergence for key frame selection [186, 187].

4.2.1 JENSEN-RÉNYI DIVERGENCE

The *Jensen-Rényi inequality* [42] is defined by

$$H_\alpha^R(\sum_{i=1}^{n} w_i p_i) - \sum_{i=1}^{n} w_i H_\alpha^R(p_i) \geq 0, \tag{4.3}$$

where

$$JR_\alpha(p_i; w_i) \equiv H_\alpha^R(\sum_{i=1}^n w_i p_i) - \sum_{i=1}^n w_i H_\alpha^R(p_i) \tag{4.4}$$

is the Jensen-Rényi divergence of the probability distributions $p_1, p_2, ..., p_n$.

A simple case of the Jensen-Shannon divergence and the Jensen-Rényi divergence is using the same weight for each probability distribution:

$$JS(p_1, p_2, \ldots, p_n) \equiv H(\frac{1}{n}\sum_{i=1}^n p_i) - \frac{1}{n}\sum_{i=1}^n H(p_i). \tag{4.5}$$

$$JR_\alpha(p_1, p_2, \ldots, p_n) \equiv H_\alpha^R(\frac{1}{n}\sum_{i=1}^n p_i) - \frac{1}{n}\sum_{i=1}^n H_\alpha^R(p_i). \tag{4.6}$$

4.2.2   THE CORE COMPUTATIONAL MECHANISM

Usually, a general video sequence can be structured in a hierachical way; that is, a video (clip) can be divided into shots and then into video frames [99]. Accordingly, the basic computational mechanism of our key frame selection method is to segment a video clip into several shots, then into subshots, and to choose key frames. The basic idea driving the segmentation of a video clip into shots/subshots is based on estimating the difference between consecutive video frames. The JS and the JR are here proposed as the metric for this purpose.

Following (4.5, 4.6) and [190], the JS-divergence and the JR-divergence values between video frames f_i and f_{i-1} [1] can be obtained:

$$JS(f_{i-1}, f_i) = H(\frac{p_{f_{i-1}} + p_{f_i}}{2}) - \frac{H(p_{f_{i-1}}) + H(p_{f_i})}{2}, \tag{4.7}$$

$$JR_\alpha(f_{i-1}, f_i) = H_\alpha^R(\frac{p_{f_{i-1}} + p_{f_i}}{2}) - \frac{H_\alpha^R(p_{f_{i-1}}) + H_\alpha^R(p_{f_i})}{2}, \tag{4.8}$$

where $p_{f_{i-1}}$ and p_{f_i} are the respective probability density functions of f_{i-1} and f_i, which are normalized from their intensity histogram distributions. Note that the difference between two video images by JS and JR is actually a sum of the correspondences for the three RGB channels.

4.2.3   LOCATING SHOTS, SUBSHOTS, AND KEY FRAMES

The procedure for our key frame selection is depicted in Figure 4.2. Now we describe it in detail. For a video clip, all the JD (here, JD refers to both JS and JR) data are obtained by evaluating JD for each pair of two consecutive video frames. Because a shot boundary reveals an abrupt variation between two consecutive video frames, we locate the shot boundaries by detecting the spikes at the JD data. If $\frac{JD(f_{i-1}, f_i)}{JD_w(f_{i-1}, f_i)} \geq \delta^*$ then a shot boundary is identified, where $JD_w(f_{i-1}, f_i)$ is an average of the $JD(f_{i-1}, f_i)$ of neighbors on a temporal window with size w (in this chapter,

---

[1]We will use in this chapter f_i to denote frame i whenever there could be a notation conflict.

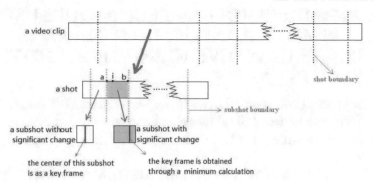

Figure 4.2: The computational mechanism for obtaining shots, subshots, and key frames. ([186] © ICPR, 2012).

$w = 5$), and 3.5 and 2.6 are respectively the thresholds experimentally defined for δ^* for JS- and JR-based approaches.

Within a shot, however, some adequate content variations could appear, for instance, possibly due to gradual scene transitions. The video content variation can be characterized by the gradient of JD, $\Delta(f_i) = JD_w(f_i, f_{i+1}) - JD_w(f_{i-1}, f_i)$, which is the rate at which a JD value changes. In practice, a window-sized version of JD gradient, $\Delta_w(f_i)$, is employed to filter out some possible minor perturbations of gradient data. If $|\Delta_w(f_I)| \geq \Delta_w^*$ (Δ_w^* is empirically taken as 1×10^{-3} and 1.5×10^{-3} for JS- and JR-based approaches, respectively), then an adequate content change inside a shot is detected at the video frame f_I. Starting from the outlier f_I, its temporally closest left and right frames f_A and f_B within this shot are obtained satisfying

$$A = max\{i \, | \, |\Delta_w(f_i)| \leq \nabla_w^* \wedge i < I\}, \tag{4.9}$$

$$B = min\{i \, | \, |\Delta_w(f_i)| \leq \nabla_w^* \wedge i > I\}, \tag{4.10}$$

here ∇_w^* is a preestablished threshold, 1×10^{-5}. Now, the video section $[A, B]$ is thus a "complete" subshot with an adequate content variation. In this way each video section with an adequate content variation is identified as a subshot, and thus a shot is divided into consecutive subshots with the borders of all the subshots with adequate content variations. If two subshots both with adequate changes are temporally close enough (namely the difference between the right border of the left subshot and the left border of the right subshot is not larger than a predefined threshold, here it is taken as 5), then the two subshots and the middle section are combined into a single subshot with an adequate content change. Note that each subshot is classified as having or not an adequate content variation.

A key frame is extracted, for a subshot, depending on the content variation. That is, the center frame of a subshot with an inadequate variation is deployed as a key frame. The key frame is obtained, for a subshot with an adequate variation, through minimizing the summed JD values between it and all the other frames.

4.3 KEY FRAME SELECTION TECHNIQUES USING TSALLIS MUTUAL INFORMATION AND JENSEN-TSALLIS DIVERGENCE FOR SHOTS WITH HARD CUTS

In this section, we present two approaches based, respectively, on Tsallis mutual information and Jensen-Tsallis divergence to detect the abrupt shot boundaries of a video sequence [178]. Then, we describe three new measures to extract the most representative keyframes.

4.3.1 MUTUAL INFORMATION-BASED SIMILARITY BETWEEN FRAMES

Cernekova et al.'s work [31] using Shannon MI between frames has been extended by using Tsallis entropy and several color spaces [178]. Specifically, the *informational frame similarity* is defined by

$$I_\alpha^{xyz}(i;j) = I_\alpha^x(i;j) + I_\alpha^y(i;j) + I_\alpha^z(i;j), \tag{4.11}$$

where the superindices x, y, and z stand for the color components in a determined color space and $I_\alpha^c(i;j)$ is the Tsallis mutual information (Equation 1.61) between frames i and j for a given color component c. As it was noted by Portes de Albuquerque et al. [134], the main motivation for the use in image and video processing of non-extensive measures, such as Tsallis entropy, is the presence of correlations between pixels of the same object in the image that can be considered as long-range correlations. Note that the use of the Tsallis generalization of mutual information will allow us to analyze the performance of the similarity measure using different entropic indices and, thus, to select the entropic index that better discriminates a shot boundary. In this chapter, we use the following color spaces: Lab (abbreviation for the CIE 1976, also called CIELAB), HSV (abbreviation for hue, saturation, and value), and RGB. Lab color space is perceptually uniform and has been designed to approximate human vision. HSV color space separates lightness from chrominance information and it is not perceptually uniform [60, 168]. Both Lab and HSV color spaces have less redundancy between the color components than RGB encoding.

The ratio IR (see Equation 4.2) is generalized by defining the *informational frame similarity ratio* as

$$IR_\alpha^{xyz}(i;i+1) = \frac{I_\alpha^{xyz}(i;i+1)}{IW_\alpha^{xyz}(i,r)}, \tag{4.12}$$

where the *average informational similarity in a window* is given by

$$IW_\alpha^{xyz}(i,r) = \frac{1}{2r} \sum_{j=i-r, j\neq i}^{i+r} I_\alpha^{xyz}(j;j+1), \tag{4.13}$$

where r is the radius of the window given by the frames $\{i-r, \ldots, i-1, i+1, \ldots, i+r\}$. In our experiments, we use $r=2$ (see Figure 4.1).

4.3.2   JENSEN-TSALLIS-BASED SIMILARITY BETWEEN FRAMES

Similar to Equation 1.25, the *Jensen–Tsallis inequality* is given by

$$JT_\alpha(\pi_1, \ldots, \pi_n; p_1, \ldots, p_n) =$$
$$H_\alpha^T(\sum_{i=1}^n \pi_i\, p_i) - \sum_{i=1}^n \pi_i\, H_\alpha^T(p_i) \geq 0, \tag{4.14}$$

where $JT_\alpha(\pi_1, \ldots, \pi_n; p_1, \ldots, p_n)$ is the *Jensen–Tsallis divergence* of probability distributions $\{p_1, \ldots, p_n\}$ with weights $\{\pi_1, \ldots, \pi_n\}$ and H_α^T is the Tsallis entropy (Equation 1.59). Since Tsallis entropy is a concave function for $\alpha > 0$, Jensen–Tsallis divergence is positive for $\alpha > 0$ [107].

Similar to the extension of mutual information (Section 4.3.1), the measure JS^{RGB} is extended using Jensen-Tsallis divergence and several color spaces.

The *Jensen–Tsallis divergence* between two frames i and j is defined as

$$
\begin{aligned}
JT_\alpha^{xyz}(i, j) &= JT_\alpha^x(\frac{1}{2}, \frac{1}{2}; p_i^x, p_j^x) \\
&+ JT_\alpha^y(\frac{1}{2}, \frac{1}{2}; p_i^y, p_j^y) \\
&+ JT_\alpha^z(\frac{1}{2}, \frac{1}{2}; p_i^z, p_j^z),
\end{aligned} \tag{4.15}
$$

where x, y, and z stand for the color components of a given color space, $JT_\alpha^c(\frac{1}{2}, \frac{1}{2}; p_i^c, p_j^c)$ is the Jensen-Tsallis divergence between the frames i and j for a given color component c (Equation 4.14), and p_i^c and p_j^c are, respectively, given by the normalized histograms of frames i and j.

As we are interested in a similarity measure between frames, we compute now the complementary measure of $JT_\alpha^{xyz}(i, j)$ with respect to its maximum value. From Equations 1.58 and 4.14, it can be seen that the maximum value of JT_α^c between two probability distributions depends on the parameter α and is given by

$$
\begin{aligned}
M_\alpha^c &= \max JT_\alpha^c(\frac{1}{2}, \frac{1}{2}; p_i^c, p_j^c) \\
&= \frac{1}{1-\alpha}(n^{1-\alpha} - 1) - \frac{1}{1-\alpha}\left(\left(\frac{n}{2}\right)^{1-\alpha} - 1\right),
\end{aligned} \tag{4.16}
$$

where $\alpha \neq 1$ and n is the number of histogram bins of the color component c. Using Jensen-Shannon divergence (Equation 1.25), it can be seen that for $\alpha = 1$, $M_\alpha^c = 1$. Thus, $M_\alpha^c - JT_\alpha^c(\frac{1}{2}, \frac{1}{2}; p_i^c, p_j^c)$ can be seen as a similarity measure between two frames for a given color component c. This allows us to define the *Jensen-Tsallis frame similarity* between two frames i and j

as

$$
\begin{aligned}
JT_\alpha^{xyz}(i,j) &= M_\alpha^x - JT_\alpha^x(\tfrac{1}{2},\tfrac{1}{2}; p_i^x, p_j^x) \\
&+ M_\alpha^y - JT_\alpha^y(\tfrac{1}{2},\tfrac{1}{2}; p_i^y, p_j^y) \\
&+ M_\alpha^z - JT_\alpha^z(\tfrac{1}{2},\tfrac{1}{2}; p_i^z, p_j^z),
\end{aligned}
\tag{4.17}
$$

where x, y, and z stand for the color components of a given color space. Observe that this measure only deals with marginal probabilities, but not with joint probabilities, and, therefore, its computation is faster than the computation of Tsallis mutual information (Equation 4.11).

Similar to the previous mutual information-based measures (Section 4.3.1), given a frame i, the *Jensen-Tsallis frame similarity ratio* between the frame similarity of pair $(i, i + 1)$ and the average frame similarity in its neighborhood is defined by

$$
JTR_\alpha^{xyz}(i, i + 1) = \frac{JT_\alpha^{xyz}(i, i + 1)}{JTW_\alpha^{xyz}(i, r)},
\tag{4.18}
$$

where

$$
JTW_\alpha^{xyz}(i, r) = \frac{1}{2r} \sum_{j=i-r, j\neq i}^{i+r} JT_\alpha^{xyz}(j, j + 1)
\tag{4.19}
$$

is the *average Jensen-Tsallis similarity in a window* of radius r. In our experiments, we use $r = 2$.

Table 4.1: List of 27 videos (with filename, number of frames, and number of shot boundaries) used in our experiments. Obtained from the video database The Open Video Project

Filename	#Frames	#Cuts	Filename	#Frames	#Cuts	Filename	#Frames	#Cuts
amc_jeep	1646	27	indi106	3610	9	parker_brothers	1595	34
apo16005	691	4	indi108	2673	13	rca_victor	1635	8
FPTVKyrgyzstan	1439	22	loop_a_lot	1754	11	sharp_calculator	1743	13
FPTVPakistan	723	18	monkey_uncle	1688	31	tide	1736	15
indi001	1686	15	newport	1765	16	trik_trak	1751	37
indi002	844	6	newport_2	1537	12	UGS07_007	3513	11
indi007	3469	24	newport_3	1538	11	uist01_13	1766	15
indi008	705	4	newport_5	1713	15	uist97_11	2856	13
indi105	1109	9	newport_8	1749	10	wth-02	385	2

4.3.3 KEYFRAME SELECTION

Given a video shot with m frames, three simple measures are proposed to extract the most representative keyframes. For a given frame i of a shot s, its average similarity with respect to the rest of frames of s can be computed using the Tsallis mutual information and the Jensen-Tsallis divergence.

The *average informational similarity* of frame i with respect to the other frames of shot s is defined by

$$AI_\alpha^{xyz}(i) = \frac{1}{m-1} \sum_{j=1, j \neq i}^{m} I_\alpha^{xyz}(i; j), \qquad (4.20)$$

where m is the number of frames of shot s, and j represents a frame of shot s different from i. From this measure, the keyframe of a shot is given by the frame with the highest average similarity. Note that AI mainly takes into account the spatial distribution of intensity values and achieves high values when the distribution in the frame i highly correlates with the distribution in the other frames of the shot (see Figure 4.3).

Similarly, the *average Jensen-Tsallis similarity* of frame i with respect to the rest of frames of shot s is defined as

$$AJT_\alpha^{xyz}(i) = \frac{1}{m-1} \sum_{j=1, j \neq i}^{m} JT_\alpha^{xyz}(i, j). \qquad (4.21)$$

Observe that AJT only compares the histograms of the frames and that the spatial distribution of the intensities is not taken into account. In this case, high values will be obtained when the histogram of a frame is similar to the histograms of the rest of the frames (see Figure 4.3).

Finally, a more global strategy is proposed to quantify the similarity between the histogram of frame i and the mean histogram of shot s. Thus, the *global Jensen-Tsallis similarity* for a frame i of shot s is defined as

$$GJT_\alpha^{xyz}(i) = JT_\alpha^{xyz}(i, \bar{s}), \qquad (4.22)$$

where $\bar{s}$ is interpreted as a virtual frame whose histogram is the average of the histograms of all frames of shot s. From this measure, the keyframe of shot s is given by the frame with maximum global similarity, that is, its histogram is the closest to the histogram that represents the whole shot (see Figure 4.4).

4.4 EXPERIMENTAL RESULTS

4.4.1 RESULTS ON JS AND JR-BASED METHODS

First, we compare the proposed algorithm based on JS with three methods: MI-based scheme [31], ETV algorithm [92], and UF technique [170]. Later, JS and UF are compared with the proposed JR-scheme. All the parameters used in competitive methods are tuned to obtain their best possible results, and in this way the numbers of key frames resulted from different methods may be slightly different. Good entropic index for JR, set by experimentation, uses values in $(0.2, 0.6)$, and is taken here as 0.4.

Extensive tests have been done for the proposed key frame selection method on a video set consisting of a lot of video sequences. These test videos are with and/or without large camera

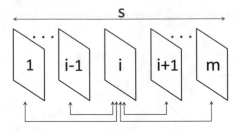

Figure 4.3: Computation scheme of the similarity between frame i and the rest of frames of shot s used by AI and AJT ([178] © Springer, 2013).

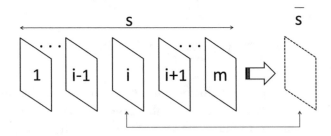

Figure 4.4: Computation scheme of the similarity between frame i and the virtual frame $\bar{s}$ used by GJT ([178] © Springer, 2013).

and object motions, as well as contain some static scenes, obtained from the web site "The Open Video Project" [1].

Test results in the JS experiment, which are obtained on a Windows PC with Intel Centrino Duo 2.0 GHz CPU and 3GB RAM, show the good performance of the JS algorithm. Table 4.2 gives a few video clips used in the tests, listing their lengths (in seconds), numbers of frames, and frame resolutions.

Experiments for JR are done based on a Windows PC with Intel Core i5 2.53 GHz CPU and 2GB RAM. Table 4.3 shows the main information of each test video, such as the time duration (in seconds) and number of video frames.

Two widely used quantitative measures, Video Sampling Error (VSE) [100] and Fidelity (FID) [35], are used for performance evaluation on the different key frame extraction methods. The similarity of two images, which is required for the computation of VSE and FID, is obtained using the weighted sum of absolute difference of momentums as introduced in [163]. A low VSE and/or a high FID is an indicator for a good performance on key frame selection, and *vice versa*.

In the JS experiment, Table 4.4 shows the quantitative measures resulted from the four methods, and clearly the JS approach does the best. Table 4.5 provides details on run time com-

Table 4.2: Test video set for the JS experiment

Video Name	Type	Length (s)	No. of Frames	Resolution
Garfield	Cartoon	16	192	240x320
Milk	Commercial	30	436	320x240
Lemon	Film Clip	30	738	320x240
Football	Sport	30	876	320x240
Metal	Film Clip	120	2880	176x144

Table 4.3: Test video set for the JR experiment

Video No.	Video Name	Length (s)	No. of Frames	Video No.	Video Name	Length (s)	No. of Frames
1	0037	27	830	11	Industry	36	1079
2	160	50	1512	12	NASAKSN-Shut	30	928
3	1234	28	854	13	senses100	58	1747
4	BOR04_002	77	2315	14	UGS06_006	40	1226
5	BOR06_002	65	1979	15	UGS08_016	87	2618
6	BOR06_004	62	1886	16	UGS13_005	25	776
7	BOR08_007	58	1759	17	cscw00_02_m4	399	8377
8	BOR14_001	36	1083	18	cscw92_05_m4	239	7159
9	BOR19_007	74	2219	19	NASAWF-GIFTS	185	5571
10	hcil2004_01_m1	36	921				

parisons. The proposed scheme can run much faster than MI-based technique, and this is due to the fact that JS is computed faster compared with MI. The UF algorithm can run faster than the other three ones, due to its very simple computation.

An example of a few key frames extracted from a test video with large video content variations by the proposed method is shown in Figure 4.5, and these key frames can represent the video content very well.

Additionally, the key frames extracted from another test video with significantly large video content variations using the four methods are depicted in Figure 4.6: the algorithm based on JS provides the most representative key frames.

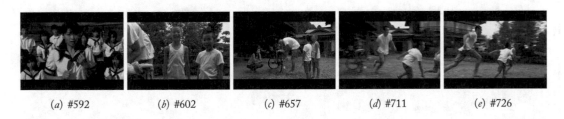

    (a) #592   (b) #602   (c) #657   (d) #711   (e) #726

Figure 4.5: The key frames selected from the video clip "Lemon" by the new proposed scheme ([187] © IEEE, 2010).

Figure 4.6: The key frames selected from the video clip "Garfield" by 4 schemes. (1)-(7): MI; (8)-(15): ETV; (16)-(23): UF; (24)-(31): JS ([187] © IEEE, 2010).

Table 4.4: Measures and numbers of key frames (NKF) by different methods in the JS experiment

		VSE	FID	NKF
Garfield	MI	124	0.249	7
	ETV	122	0.241	8
	UF	130	0.246	8
	JS	90	0.293	8
Milk	MI	103	0.260	11
	ETV	104	0.250	13
	UF	180	0.247	13
	JS	73	0.364	13
Lemon	MI	94	0.437	23
	ETV	92	0.436	22
	UF	127	0.393	22
	JS	68	0.438	22
Football	MI	485	0.370	25
	ETV	487	0.370	27
	UF	542	0.367	27
	JS	480	0.370	27
Metal	MI	1323	0.288	23
	ETV	1393	0.283	21
	UF	1480	0.236	21
	JS	929	0.305	21

Table 4.5: Runtime by different methods (s) in the JS experiment

	Garfield	Milk	Lemon	Football	Metal
MI	11.3	42.2	74.1	73.8	99.1
ETV	15.9	73.3	138.1	108.8	376.9
UF	0.4	1.7	3.4	3.4	3.7
JS	1.1	3.1	5.4	5.5	8.7

As for the JR experiment, in Figure 4.7, the VSE, FID, and run time values, which result from the different algorithms, are displayed.

Figures 4.8–4.11 provide several examples to show the better performance by JR. The JR and JS plots for the two test videos "UGS06_006" and "BOR14_001" are given in Figures 4.8 and 4.10: we use a black dash-dotted line, a green dashed line, and a red point to, respectively, represent a shot boundary, a subshot separation and a key frame. Correspondingly, the key frames extracted are shown in Figures 4.9 and 4.11.

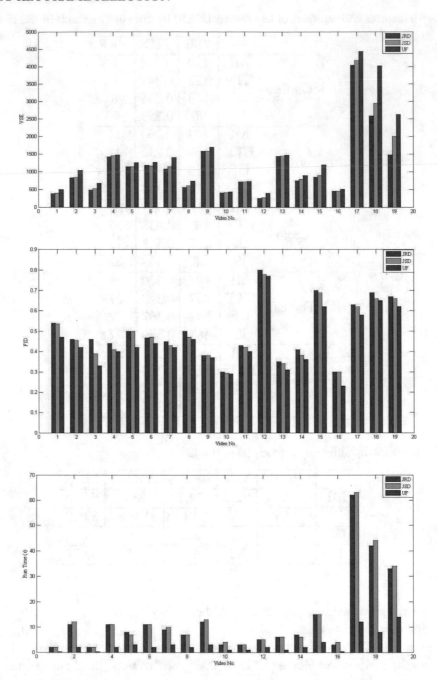

Figure 4.7: The VSE, FID, and run time by different algorithms on the test videos in the JR experiment ([186] © ICPR, 2012).

Figures 4.8 and 4.9 exhibit the behaviors of different key frame selection methods, for a clip of "UGS06_006." This video clip is with big content changes, including a substantial camera motion (#695–#865) and panning (#866–#929), a large object moving (#930–#991) and a static scene with a caption that is being generated (#992–#1226). JR technique indicates a shot boundary at #765, thus two key frames #730 and #815 are extracted and they can satisfactorily point out that the camera is moving. In addition, the key frame #960, resulted from the identification of a shot cut at #991 by JR, clearly renders an object motion. Unfortunately, the key frames selected by JS cannot well depict the camera and object motions. In summary, our JR-based technique does very well for identifying key frames.

Figures 4.10 and 4.11 show that JR achieves superior results than JS on "BOR14_001," which is with abundant scene switches and is a typical general video sequence in many real applications. JR can extract the key frames indicating scene switches (#250 and #346), and this obviously benefits the understanding of the original video. The key frames by JR have less redundancy, however JS selects the duplicated #889 and #976 as key frames. As for the image quality of the extracted key frames themselves, the obtained by JR are usually acceptable, which is an advantage for a key frame selection technique.

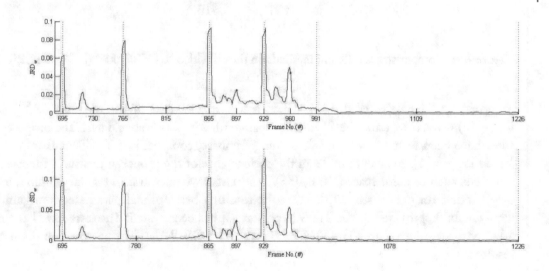

Figure 4.8: The JR and JS plots for the test video "UGS06_006" ([186] © ICPR, 2012).

4.4.2 RESULTS ON JT AND TMI DRIVEN TECHNIQUES

In this section it is analyzed the performance of the proposed Tsallis entropy-based measures, using different color spaces and histogram binnings, to deal with shot boundary detection and keyframe selection. These measures are compared with the mutual information-based measure

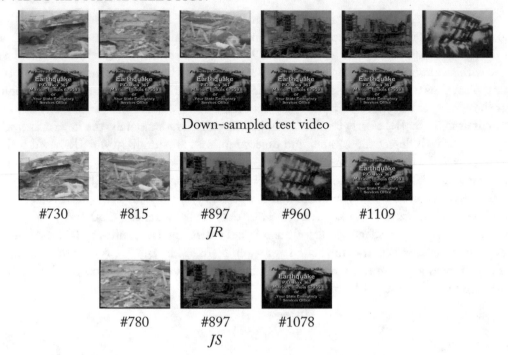

Figure 4.9: Comparison of different methods on the test video "UGS06_006" ([186] © ICPR, 2012).

proposed by Cernekova et al. [31], which is a particular case of the measure IR_α^{xyz} (Equation 4.12) when $\alpha = 1$ and the RGB color space with 256 bins is considered. The proposed measures have been tested first against a training database, composed of 27 videos from the Open Video Project [1], to analyze in detail the performance of the proposed measures for several parameters, such as color space (RGB, HSV, and Lab have been used), regular binning, and entropic index; for details see [178]. With the resulting best optimal parameter configurations a large database, provided by the TrecVid project [2] has been tested. The tests have been run on a PC with an Intel© Core™i5 430M 2.27GHz and 4 GB RAM. The next section discusses the results.

Testing Database

In this section, a testing database is used to compare the proposed measures in a real environment. This large database is provided by the TrecVid project [2] and it contains 17 videos with resolution of 352×288 pixels and a total duration of 7 h and 29 min (see Table 4.6, where the number of frames and the number of shot boundaries for each video are also shown). These videos were used to test several methods in the shot boundary detection task in TrecVid 2007. The results obtained by these methods are available in [3]. The shot boundary positions are given by the ground truth

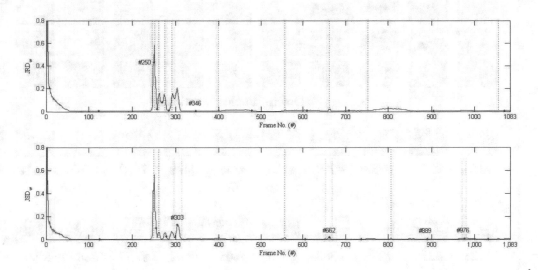

Figure 4.10: The JR and JS plots for the test video "BOR14_001" ([186] © ICPR, 2012).

provided by the TrecVid project together with the video database. The gradual transitions have not been considered in our experiments.

Table 4.6: List of 17 videos (with filename, number of frames, and number of shot boundaries) used in our experiments. Obtained from the TrecVid project [2]

Filename	#Frames	#Cuts	Filename	#Frames	#Cuts
BG_2408	35892	101	BG_9401	50049	89
BG_11362	16416	104	BG_14213	83115	106
BG_34901	34389	224	BG_35050	36999	98
BG_35187	29025	135	BG_36028	44991	87
BG_36182	29610	95	BG_36506	15210	77
BG_36537	50004	259	BG_36628	56564	192
BG_37359	28908	164	BG_37417	23004	76
BG_37822	21960	119	BG_37879	29019	95
BG_38150	52650	215			

Table 4.7 summarizes the results obtained with the proposed measures applied to the testing database. In this experiment, we only analyze the Shannon-based measures IR_1^{RGB} and JTR_1^{RGB} and the measures that have achieved the best results with the training database. A shot boundary

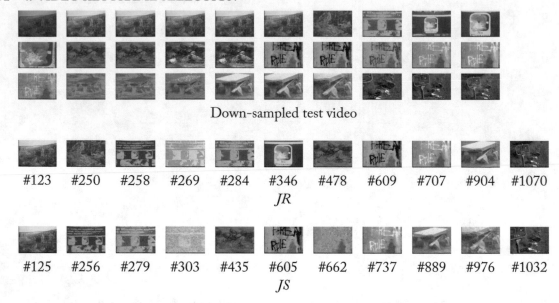

Down-sampled test video

| #123 | #250 | #258 | #269 | #284 | #346 | #478 | #609 | #707 | #904 | #1070 |

JR

| #125 | #256 | #279 | #303 | #435 | #605 | #662 | #737 | #889 | #976 | #1032 |

JS

Figure 4.11: Comparison of different methods on the test video "BOR14_001" ([186] © ICPR, 2012).

is detected when a pair of frames has a similarity value lower than a given threshold. The obtained results are compared with the ground truth.

To evaluate the results, the measures *precision* and *recall* are used. The precision is defined as

$$P = \frac{TP}{TP + FP}, \tag{4.23}$$

where TP is the number of true positives (i.e., the shot boundaries that the algorithm detects and that correspond to the real ones according to the ground truth) and FP is the number of false positives (i.e., the shot boundaries that the algorithm detects and that do not correspond to the real ones according to the ground truth). The recall is defined as

$$R = \frac{TP}{TP + FN}, \tag{4.24}$$

where FN is the number of false negatives (i.e., the shot boundaries that the algorithm does not detect and that correspond to the real ones according to the ground truth). The measures precision and recall take values in the range $[0, 1]$, being 1 the best value. In addition to the previously defined measures precision and recall, the harmonic mean of both measures, also called *F-measure*, is used as a single value that summarizes both precision and recall. The F-measure is defined as

$$F = \frac{2}{\frac{1}{P} + \frac{1}{R}}.$$

(4.25)

As it can be seen in Table 4.7, the best results are obtained with $IR_{1.7}^{HSV}$ and 8 bins using a threshold value of 0.2. Note also that the use of the Tsallis generalization notably improves the results (F-measure varies from 0.9384 in the second row to 0.9495 in the forth row), and also the reduction of histogram bins and the use of the HSV color space have a great impact on the results (F-measure varies from 0.8558 in the first row to 0.9384 in the second row). If the results are analyzed in a detailed way, we observe that approximately half of the errors (50 of 104) come from a single video, BG_36628, and are due to a scene with black frames corrupted by noise. As these frames contain very little information, the proposed measure does not perform properly because the normalization makes it very sensitive to the presence of noise. This problem could be easily solved by, for instance, not taking into account frames with low information (i.e., low entropy value), similarly to the strategy proposed by Cernekova et al. [31] to detect fades. The JTR-based measures obtain in general worse results than the IR-based ones, but note that $JTR_{0.5}^{Lab}$ with 128 histogram bins obtain clearly better results than the measures IR_1^{RGB} and JTR_1^{RGB}).

Table 4.7: Results obtained using the testing database and a threshold value as stopping criterion

Measure	Color variables	#Bins	Threshold	Precision	Recall	F-measure
IR_1	$R \oplus G \oplus B$	256	0.4	0.7811	0.9463	0.8558
IR_1	$H \oplus S \oplus V$	8	0.2	0.9542	0.9231	0.9384
$IR_{1.5}$	$H \oplus S \oplus V$	8	0.2	0.9565	0.9352	0.9457
$IR_{1.7}$	$H \oplus S \oplus V$	8	0.2	0.9531	0.9459	0.9495
JTR_1	$R \oplus G \oplus B$	256	0.94	0.8664	0.8936	0.8798
JTR_1	$L \oplus a \oplus b$	32	0.98	0.7793	0.9696	0.8641
$JTR_{0.5}$	$L \oplus a \oplus b$	128	0.98	0.8460	0.9584	0.8987

Table 4.8 shows the results obtained with the same measures than the previous experiment when the number of cuts is a priori known and used as a stopping criterion. These results can be seen as the best possible achievable results when the best particular threshold is selected for each video. Note that, in this case, precision and recall take the same value, since FP is equal to FN and, thus, the F-measure (i.e., the harmonic mean) also takes this value. The general behaviour of the measures is similar to the previous experiment, obtaining the best results when the perceptual color spaces and the generalized entropies are used. There are some minor differences, such as the better behaviour of $IR_{1.5}^{HSV}$ with respect to the one of $IR_{1.7}^{HSV}$. From this fact, it can be established that $IR_{1.5}^{HSV}$ has more capacity to obtain good results, while $IR_{1.7}^{HSV}$ is a more stable measure to be used with a single threshold value for all the videos.

Table 4.8: Results obtained using the testing database and the number of cuts as stopping criterion

Measure	Color variables	#Bins	F-measure
IR_1	$R \oplus G \oplus B$	256	0.9186
IR_1	$H \oplus S \oplus V$	8	0.9495
$IR_{1.5}$	$H \oplus S \oplus V$	8	0.9562
$IR_{1.7}$	$H \oplus S \oplus V$	8	0.9548
JTR_1	$R \oplus G \oplus B$	256	0.8855
JTR_1	$L \oplus a \oplus b$	32	0.9052
$JTR_{0.5}$	$L \oplus a \oplus b$	128	0.9047

Another important aspect to be considered is the computation time. Figure 4.12 plots the F-measure versus the computation time per frame in milliseconds spent on detecting the shot boundaries for the measures IR_1^{RGB} with 256 bins, IR_1^{HSV} with 8 bins, $IR_{1.7}^{HSV}$ with 8 bins, JTR_1^{RGB} with 256 bins, JTR_1^{Lab} with 32 bins, and $JTR_{0.5}^{Lab}$ with 128 bins. As we could expect, the computational time is very sensitive to the number of bins. On the other hand, the difference between the Shannon entropy-based measures and their extension based on Tsallis entropy is not significant (especially, for a low number of bins). It is also important to notice that the computational cost of the measures based on Jensen-Tsallis divergence is between two times and four times lower than the measures based on mutual information, since in the first case the computation of the joint histogram between frames is not required.

As a conclusion, we have seen that IR with $\alpha = 1.7$ and perceptually aimed color spaces achieves the best performance, significantly improving the performance of the mutual information-based measure proposed by Cernekova et al. We have also seen that JTR, with $\alpha = 0.5$, 128 bins, and $L \oplus a \oplus b$ color variables, obtains a good tradeoff between accuracy and computational cost.

4.5 CONCLUSION

In this chapter, several information-theoretic approaches for video shot boundary and key frame selection have been presented in order to understand a video clip effectively and efficiently. From the initial work of Cernekova et al. [31] new algorithms have been developed [178, 186, 187].

Some algorithms extract key frames of a video clip using Jensen-Shannon divergence and Jensen-Rényi divergence for evaluating differences between successive video frames, and utilize Jensen-Shannon and Jensen-Rényi gradient values for locating video sections with large video content variations. The performance of these schemes is justified by extensive experimental results and by the comparisons with classic and close related approaches.

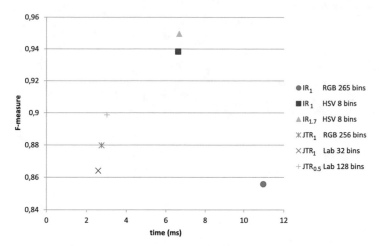

Figure 4.12:   F-measure and computation time per frame in milliseconds, for the measures analyzed in Table 4.7 ([178] © Springer, 2013).

The other approaches are based on Tsallis mutual information and Jensen-Tsallis divergence to deal with video shot boundary detection and keyframe selection. Their discriminatory capacity has been analyzed for several color spaces (RGB, HSV, and Lab), regular binnings, and entropic indices. Different experiments have shown that the optimal capacity is obtained by the Tsallis mutual information similarity measure using HSV and Lab color spaces. In general, the reduction of the number of histogram bins also improves the results.

CHAPTER 5

Informational Aesthetics Measures

5.1 INTRODUCTION

In 1928, George D. Birkhoff formalized the *aesthetic measure* of an object as the quotient between *order* and *complexity* [19]. From Birkhoff's work, Bense [16], together with Abraham Moles, developed the *Informational Aesthetics*[1] where the concepts of order and complexity were defined from the notion of *information* provided by Shannon's work [42]. As Birkhoff stated, it is very difficult to formalize those concepts which are dependent on the context, author, observer, etc. Scha and Bod [154] claimed that in spite of the simplicity of these beauty measures, "if we integrate them with other ideas from perceptual psychology and computational linguistics, they may in fact constitute a starting point for the development of more adequate formal models." Bense proposed a general schema which establishes that the artistic production is characterized by the transition from the initial repertoire to the final product. In the creative process, order is generally produced from disorder. Bense assigned a complexity to the repertoire or palette and an order to the distribution of the palette elements on the artistic product.

In this chapter, we present a set of measures that conceptualize the Birkhoff's aesthetic measure from an informational point of view [140]. A first group of global measures, based on Shannon entropy and Kolmogorov complexity, gives us a scalar value associated with an artistic object. A second group of compositional measures extends the previous analysis in order to capture the structural information of the object. In particular, an information channel that fits well with the creative channel presented by Bense is introduced.

All the measures introduced describe complementary aspects of the aesthetics experience and are normalized for an easier comparison. The behavior of these measures is shown using three sets of paintings representing different styles which cover a representative range from randomness to order. Our experiments show that both global and compositional measures well extend the Birkhoff's measure and help us to understand and quantify the creative process.

After the presentation of the aesthetic measures and assuming a classification of van Gogh's paintings in six periods [22], we propose an informational dialogue with van Gogh's artwork showing a significant consistency between the proposed aesthetic measures and period-styles, and also providing new tools to study the role of color in the painting composition [142]. A subset of

---

[1] We keep the name used in English literature on the subject although we believe a more adequate name would be *information-theoretic aesthetics* according to the original German term.

van Gogh's paintings is analyzed using three tools: the entropy of the palette, the compressibility of the image, and an information channel to capture the basic structure of the painting.

To investigate further, we also study —using the full set of color digital images of van Gogh's paintings available in *The Vincent van Gogh Gallery* of David Brooks [22]—whether key features of van Gogh periods can be determined by an extended set of informational measures [143]. We will focus mainly on van Gogh's Auvers period and will try to investigate whether several information-theoretic measures can support the claim of art critics on his evolution of palette and composition. We will also study how far van Gogh's last period was from his other periods, and try to trace his artistic development. To this end, we will employ the previously defined measures together with a set of measures that take into account spatial information. In addition, we will use a visual tool to analyze the palette.

Finally, we show two measures which quantify the information associated with each color and region of a painting. These measures permit us to visualize the most informative or salient colors and elements (objects or regions) of an image. The goal of this van Gogh's analysis is to show how a set of tools could help to discover relevant characteristics of a painting (or painter's style) which could go unnoticed by the observer.

5.2 ORIGINS AND RELATED WORK

In 1928, Birkhoff formalized the notion of beauty by the introduction of the aesthetic measure, defined as the ratio between *order* and *complexity* [19], where "the complexity is roughly the number of elements that the image consists of and the order is a measure for the number of regularities found in the image" [154]. Birkhoff suggested that aesthetic feelings stem from the harmonious interrelations inside the object and that the aesthetic measure is determined by the order relations in the object. Birkhoff understood the impossibility of comparing objects of different classes and accepted that the aesthetic experience depends on each observer. Hence, he proposed restricting the group of observers and to apply only the measure to similar objects.

According to Birkhoff, the aesthetic experience is based on three successive phases:

1. a preliminary effort of attention, which is necessary for the act of perception and increases proportionally to the *complexity* (C) of the object;

2. the feeling of value or *aesthetic measure* (M) which comes from this effort; and

3. the verification that the object is characterized by certain harmony, symmetry or *order* (O), which seems to be necessary for the aesthetic effect.

From these considerations, Birkhoff defined the aesthetic measure as $M = \frac{O}{C}$. Later, in 1965, Bense [16] and Moles [110] interpreted Birkhoff's measure from an information-theoretic perspective.

Using information theory, Bense [16] proposed both the *redundancy* and Shannon entropy to quantify, respectively, the order and the complexity. According to Bense, in any artistic process of creation, we have a determined *repertoire* of elements (such as colors, sounds, phonemes, etc.) which is *transmitted* to the final *product*. The creative process is a selective process (i.e., to create means to select). For instance, if the repertoire is given by a palette of colors with a probability distribution, the final product (in our case, a painting) is a selection (a realization) of this palette on a canvas. While the distribution of elements of an aesthetic state has a certain *order*, the repertoire shows a certain *complexity*. Bense also distinguished between a global complexity, formed by partial complexities, and a global order, formed by partial orders. His contemporary Moles [110] considered order expressed not only as redundancy but also as the degree of predictability.

Other authors have also introduced several measures with the purpose of quantifying aesthetics. Koshelev et al. [86] considered that the running time $t(p)$ of a program p which generates a given design is a formalization of Birkhoff's complexity C, and a monotonically decreasing function of the length of the program $l(p)$ (i.e., Kolmogorov complexity) represents Birkhoff's order O. Thus, looking for the most attractive design, the aesthetic measure is defined by $M = 2^{-l(p)}/t(p)$. Machado and Cardoso [103] established that an aesthetic visual measure depends on the ratio between *image complexity* and *processing complexity*. Both are estimated using real-world compressors (jpg and fractal, respectively). They considered that images that are simultaneously visually complex and easy to process are the images that have a higher aesthetic value. Excellent overviews of the history of the aesthetic measures can be found in the surveys by Greenfield [62] and Hoenig [72].

Next, we review Rigau et al. [141, 142, 143], where a set of information-theoretic measures were presented to study some informational aspects of a painting related to its palette and composition. Some of these measures are used to discriminate different painting styles and to analyze the evolution of van Gogh's artwork.

5.3 GLOBAL AESTHETIC MEASURES

From the creative process proposed by Bense, we consider three basic concepts: initial repertoire, used palette, and final color distribution. The initial repertoire is given by the basic states (in our case, a wide range of colors which we assume as finite and discrete). The palette (selected repertoire) is the range of colors selected by the artist with a given probability distribution. From the palette, the artist distributes or arranges the colors on a physical support (canvas) obtaining the final product. Next, we present a set of measures that use these concepts to extend Birkhoff's aesthetic measure.

For a given color image $\mathcal{I}$ of N pixels, we use an sRGB color representation[2] based on a repertoire (i.e., an alphabet $\mathcal{X}_{rgb}$) of 256^3 colors. Note that any other color system could be used. The range of $\mathcal{X}_{rgb}$ can be reduced using the luminance[3] Y_{709} (i.e., $\mathcal{X}_\ell = [0, 255]$). From the

---

[2]A tristimulus color system of 256 discrete values for each channel (red, green, and blue). Each channel has 8 bits of information and, therefore, a color depth of 24 bits per pixel is used (i.e., 256^3 possibilities).

[3]The luminance is a measure of the density of luminous intensity of a pixel computed as a lineal combination of its RGB channels. We use the Rec. 709: $Y = 0.212671R + 0.715160G + 0.072169B$.

normalization of the intensity histograms of $\mathcal{X}_{rgb}$ and $\mathcal{X}_\ell$, using 256^3 (B_{rgb}) and 256 (B_ℓ) bins, respectively, the probability distributions of the random variables X_{rgb} and X_ℓ are obtained. The maximum entropy H_{max} for these random variables is $\log|B_{rgb}| = 24$ and $\log|B_\ell| = 8$, respectively (see Section 1.1). Throughout this chapter, the following notions are used:

– palette (X_{rgb} or X_ℓ), given by the normalized intensity histogram of the image;

– entropy of the palette or uncertainty of a pixel, $H(C)$, obtained from $H(X_{rgb})$ or $H(X_\ell)$;

– information content or uncertainty of an image ($N \times H(C)$); and

– Kolmogorov complexity of an image (K).

These concepts are applied to a set of paintings of Mondrian (Composition with Red, 1938–1939 [111], Composition with Red, Blue, Black, Yellow, and Gray, 1921 [112], and Composition with Grid 1, 1918 [113]), Seurat (Figure 5.1 a), and van Gogh (Figure 5.1 b). Their sizes are given in Table 5.1, where we have also indicated the size and compression ratio achieved by the jpg compressor.

Table 5.1:   For the test paintings of Mondrian, Seurat, and van Gogh, we give the size (pixels and bytes) of the original files, and their size (bytes) and compression ratio using jpg compression with the maximum quality option

Image				jpg	
Image	Painting	Pixels	Bytes	Bytes	Ratio
[111]	Mondrian-1	316888	951862	160557	5.928
[112]	Mondrian-2	139050	417654	41539	10.055
[113]	Mondrian-3	817740	2453274	855074	2.869
5.1.(a.1)	Seurat-1	844778	2535422	1473336	1.721
5.1.(a.2)	Seurat-2	857540	2572674	1530889	1.681
5.1.(a.3)	Seurat-3	375750	1128306	519783	2.171
5.1.(b.1)	VanGogh-1	831416	2495126	919913	2.712
5.1.(b.2)	VanGogh-2	836991	2511850	862274	2.913
5.1.(b.3)	VanGogh-3	856449	2570034	1203527	2.135

5.3.1   SHANNON'S PERSPECTIVE

From Equation 1.2, the entropy $H(C)$ of a random variable C taking values c in $\mathcal{C}$ with distribution $p(c) = Pr\{C = c\}$ is defined by

$$H(C) = -\sum_{c \in \mathcal{C}} p(c) \log p(c). \qquad (5.1)$$

(*a.1*) The Seine at Le Grande Jatte
Georges-Pierre Seurat, 1888

(*a.2*) Forest at Pontaubert
Georges-Pierre Seurat, 1881

(*a.3*) Sunday Afternoon on the Island of La Grande Jatte
Georges-Pierre Seurat, 1884-1886

(*b.1*) The Starry Night
Vincent van Gogh, 1889

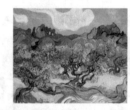

(*b.2*) Olive Trees with the Alpilles in the Background
Vincent van Gogh, 1889

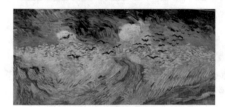

(*b.3*) Wheat Field Under Threatening Skies
Vincent van Gogh, 1890

Figure 5.1: Set of paintings used in the tests.

The palette entropy $H(C)$ fulfills $0 \leq H(C) \leq H_{max}$ and can be interpreted as the average uncertainty of a pixel color.

As we have seen in Section 5.2, Bense proposed to use the redundancy to measure the *order* in an aesthetic object. Applying this idea to an image or painting, the *absolute redundancy* $H_{max} - H(C)$ expresses the reduction of uncertainty due to the choice of a palette with a given color probability distribution instead of taking a uniform one. Thus, the aesthetic measure can be expressed as the *relative redundancy*:

$$M_B = \frac{H_{max} - H(C)}{H_{max}},$$ (5.2)

where M_B takes values in $[0, 1]$ and expresses the reduction of pixel uncertainty due to the choice of a palette with a given color probability distribution instead of a uniform distribution [141]. In our tests, M_B has been computed from alphabet $\mathcal{X}_{rgb}$. From a coding perspective, this measure represents the gain obtained using an optimal code to compress the image (Equation 1.35). The redundancy expresses one aspect of the creative process: the selection of the palette used by the artist.

Table 5.2 shows that there are significative differences of values of M_B for the set of test paintings, where the entropy of a pixel has been computed using $H(C) = H(X_{rgb})$, where $H_{max} = 24$. From Mondrian-1 to VanGogh-3, the results are gradually lower. This is mainly due to a high color homogeneity in Mondrian's paintings and a major color diversity in Seurat's and Van Gogh's ones. Note that this measure only reflects the palette information but does not take into account the spatial distribution of colors on canvas. Thus, the geometry (Mondrian), pointillism's randomness (Seurat), and landscape elements (van Gogh and Seurat) are compositional features perceived by a human observer but not captured by M_B. The following measures take into account this spatial feature.

5.3.2   KOLMOGOROV'S PERSPECTIVE

From a Kolmogorov complexity perspective, the *order* in an image can be measured by the difference between the image size (obtained using a constant length code for each color) and its Kolmogorov complexity K (Section 1.11). This corresponds to the space saving defined as the reduction in size relative to the uncompressed size. The normalization of the order gives us the aesthetic measure

$$M_K = \frac{N \times H_{max} - K}{N \times H_{max}},$$ (5.3)

where M_K takes values in $[0,1]$ and expresses the degree of order of the image without any a priori knowledge on the palette (the higher the order of the image, the higher the compression ratio). Due to the non-computability of K [97], real-world compressors are used to estimate it (i.e., the value of K is approximated by the size of the corresponding compressed file). Observe that a compressor takes advantage of both the degree of order of the selected palette and the color position in the canvas. We selected the jpg compressor due to its ability to discover patterns, in spite of (or thanks to) losing information not perceived by the human eye. This is closer to the aesthetic experience than using lossless compressors which usually have lower compression ratios to keep all the original information, including that which cannot be distinguished by a human observer. Nevertheless, to avoid losing significative information, we use a jpg compressor with the maximum quality option (see Table 5.1).

In Table 5.2, M_K has been calculated using $H_{max} = 24$. Although a strict ordering upon M_K values mixes paintings of different artists, the averages of the three sets of paintings are clearly separated. In descending order, the groups are Mondrian, Van Gogh, and Seurat. Observe that the pairs of paintings (Mondrian-3, VanGogh-2) and (VanGogh-3, Seurat-1) have very similar M_K values. This is probably due to the fact that the compressor is able to detect more homogeneity (or heterogeneity) than the human eye. For instance, the interior of some regions in the Mondrian-3 painting is more heterogeneous than would appear at first glance.

Frieder Nake, disciple of Bense and one of the pioneers of *computer or algorithmic art* (i.e., art explicitly generated by an algorithm), considered a painting as a hierarchy of signs, where at each level of the hierarchy the statistical information content could be determined. He conceived the computer as a *Universal Picture Generator* capable of "creating every possible picture out of a combination of available picture elements and colors [116]." We can see that Nake's theory of algorithmic art fits well with Kolmogorov's perspective, since the Kolmogorov complexity of a painting can be considered as the length of the shortest program generating it.

5.3.3   ZUREK'S PERSPECTIVE

Looking at a *system* from an observer's angle, Zurek [192] defined the *physical entropy* as the sum of the missing information (Shannon entropy) and the algorithmic information content (Kolmogorov complexity) of the available data:

$$S_d = H(X_d) + K(d), \tag{5.4}$$

where d is the observed data of the system, $K(d)$ is the Kolmogorov complexity of d, and $H(X_d)$ is the conditional Shannon entropy or our ignorance about the system given d.

Physical entropy reflects the fact that measurements increase our knowledge about a system. In the beginning, we have no knowledge about the state of the system, therefore the physical entropy reduces to the Shannon entropy, reflecting our total ignorance. If the system is in a regular state, physical entropy decreases with the more measurements we make. In this case, we increase our knowledge about the system and we may be able to compress the data efficiently. If the state is not regular, then we cannot achieve compression and the physical entropy remains high. According to Zurek, this compression process can be seen from the perspective of an Information

Gathering and Using System (IGUS), like a *Maxwell's demon*,[4] entity capable of performing measurements and of modifying its strategies on the basis of the outcomes of the measurements.

We show another version of Birkhoff's measure based on Zurek's physical entropy. Zurek's work permits us to look at the creative process as an evolutionary process from the initial uncertainty (Shannon entropy) to the final order (Kolmogorov complexity). This approach can be interpreted as a transformation of the initial *probability distribution* of the palette of colors to the *algorithm* which describes the final painting.

Inspired by physical entropy (Equation 5.4), we define a measure given by the ratio between the *reduction of uncertainty* (due to the compression achieved by Kolmogorov complexity) and the initial *information content* of the image. Assuming that the information content of an image is given by the Shannon entropy of each pixel times the number of pixels ($N \times H(C)$), we have

$$M_Z = \frac{N \times H(C) - K}{N \times H(C)}.$$ (5.5)

This normalized ratio quantifies the degree of order created from a given palette.

In Table 5.2, M_Z has been computed using the jpg compressor, $H(C) = H(X_{rgb})$, and $H_{max} = 24$. Taking the average of M_Z for each artist, we observe that they have been ordered in the same way as in the previous measure M_K. The low values for Seurat's paintings are due to their low compression ratio because of the pointillist style (see Table 5.1).

Table 5.2: Entropy $H(X_{rgb}$ and global aesthetic measures M_B, M_K, and M_Z for the test paintings

Painting	$H(X_{rgb})$	M_B	M_K	M_Z
Mondrian-1	8.168	0.660	0.831	0.504
Mondrian-2	9.856	0.589	0.900	0.758
Mondrian-3	14.384	0.401	0.651	0.418
Seurat-1	14.976	0.376	0.419	0.068
Seurat-2	18.180	0.243	0.405	0.214
Seurat-3	17.045	0.290	0.539	0.351
VanGogh-1	17.204	0.283	0.631	0.485
VanGogh-2	17.288	0.280	0.657	0.523
VanGogh-3	17.689	0.263	0.532	0.364

The plots shown in Figure 5.2 express, for three paintings, the *evolution* of the physical entropy as we take more and more measurements. To simulate this, the content of each painting is progressively discovered (by columns from left to right), reducing the missing information (Shannon entropy) and compressing the discovered one (Kolmogorov complexity). Mondrian paintings show on average a greater order than those of Van Gogh, and Van Gogh more than those

---

[4]The Maxwell's demon was presented in 1867 by the physicist James Clerk Maxwell as an "intelligent being" capable of splitting a set of molecules of gas into fast and slow ones, apparently violating the second law of thermodynamics.

of Seurat. Thus, our progressive knowledge about the painting can be more efficiently compressed or comprehended in the Mondrian case than in the other cases.

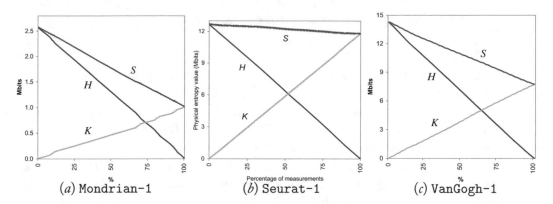

(a) Mondrian-1   (b) Seurat-1   (c) VanGogh-1

Figure 5.2: The evolution of physical entropy (S) (missing information H + Kolmogorov complexity K) for three of the test paintings. The missing information is captured by $H(C) = H(X_{rgb})$ and the Kolmogorov complexity has been approximated using the jpg compressor. The percentage of measurements from the painting and the value of the physical entropy in Mbits are shown on the x-axis and y-axis, respectively ([141] © IEEE, 2008).

The global measures analyzed above can be understood from the complexity of the initial repertoire (logarithm of the number of states of the repertoire), the complexity of the selected palette (Shannon entropy), and the complexity of the final distribution (Kolmogorov complexity). From these complexities, the order is obtained measuring the differences between them. Thus, in M_B, $H_{max} - H(C)$ is the redundancy of the palette, in M_K, $N \times H_{max} - K$ is the compression achieved due to the order in the product, and in M_Z, $N \times H(C) - K$ is the reduction of uncertainty which is produced in the process of observing or recognizing the final product. These differences quantify the creative process: the first represents the selection process from the initial repertoire, the second captures the order in the distribution of colors, and the third expresses the transition between the palette and the artistic object.

5.4   COMPOSITIONAL AESTHETIC MEASURES

As we have seen, Bense considered the creative act as a transition process from an initial repertoire to the distribution of its elements on the physical support. In this section, our interest is focused on the introduction of measures to analyze the composition of an image (i.e., the spatial distribution of colors from a given palette) [141].

5.4.1   ORDER AS SELF-SIMILARITY

In order to analyze the composition of an image, the measures used will have to quantify the degree of correlation or similarity between the parts of the image. The Jensen-Shannon divergence and the similarity metric are appropriate to capture the spatial order.

Shannon's Perspective

From Shannon's point of view, the similarity between the different parts of an image can be computed using the Jensen-Shannon divergence Equation 1.25, which is a measure of discrimination between probability distributions. In our case, this divergence can be used to calculate the dissimilarity between the intensity histograms of diverse regions. Thus, for a given decomposition of an image, the Jensen-Shannon divergence will quantify the spatial heterogeneity.

While the ratio between the JS-divergence and the initial uncertainty $H(C)$ of the image expresses the degree of *dissimilarity*, the complementary of this ratio is defined as a measure of *self-similarity*:

$$
\begin{aligned}
M_j(n) &= 1 - \frac{JS(\pi_1, \ldots, \pi_n; p_1, \ldots, p_n)}{H(C)} \\
&= 1 - \frac{H(C) - \sum_{i=1}^{m} \pi_i H(p_i)}{H(C)} \\
&= \frac{\sum_{i=1}^{m} \pi_i H(p_i)}{H(C)},
\end{aligned}
\tag{5.6}
$$

where n is the resolution level (i.e., the number of regions desired), π_i is the normalized area of region i, p_i represents the probability distribution of region i, and $H(p_i)$ is its entropy. The self-similarity measure takes values in $[0,1]$, decreasing its value with a finer partition. For a random image and a coarse resolution, it should be close to 1.

Table 5.3 shows the values of M_j for the set of paintings. In our tests, the paintings have been decomposed in a 4×4 regular grid and the histograms have been computed using the luminance Y_{709}. Observe how the very high visual similarity between the parts of a Seurat painting fits with the high values of M_j. On the other hand, the lower self-similarity of `Mondrian-2` is due to the presence of regions with very different colors.

Kolmogorov's Perspective

To measure the similarity between two parts of an image, we can now use Equation 1.64, the normalized information distance (NID). As we have seen in Section 1.11, the information distance between two subimages is the length of the shortest program that is needed to transform the two subimages into each other. If we consider the degree of order of an image as its self-similarity, this can be measured from the average of NID between each pair of subimages:

$$
M_k(n) = 1 - \text{avg}_{1 \le i < j \le n} \{NID(i, j)\},
\tag{5.7}
$$

where n is the number of regions or subimages provided by a given decomposition, and $NID(i, j)$ the distance between the subimages $\mathcal{I}_i$ and $\mathcal{I}_j$. This value ranges from 0–1 and expresses the degree of order inside the image.

In Table 5.3, the values of M_k for the set of paintings are calculated using a 4×4 regular grid and $NCD(i, j)$ Equation 1.65 as an approximation of $NID(i, j)$. For our specific case, the values of $C(\mathcal{I}_i)$ and $C(\mathcal{I}_i, \mathcal{I}_j)$ in NCD have been computed ignoring the rest of the canvas information (i.e., zero luminance in $\mathcal{I} - \mathcal{I}_i$ and $\mathcal{I} - \mathcal{I}_i - \mathcal{I}_j$, respectively). Similar to the previous compositional measure M_j, the paintings are classified according to the artist. However, note that the order is reversed. This is because, while M_j only measures the similarity between the palettes of the regions, M_k also measures the similarity of the spatial distribution of the palettes on the canvas.

Table 5.3: The compositional aesthetic measures M_j, M_k, and M_s for the set of test paintings computed for $n = 16$

Painting	$H(X_\ell)$	M_j	M_k	M_s
Mondrian-1	5.069	0.900	0.312	0.166
Mondrian-2	6.461	0.762	0.335	0.352
Mondrian-3	7.328	0.969	0.198	0.060
Seurat-1	7.176	0.984	0.161	0.025
Seurat-2	7.706	0.979	0.147	0.032
Seurat-3	7.899	0.960	0.164	0.055
VanGogh-1	7.858	0.953	0.179	0.070
VanGogh-2	7.787	0.948	0.170	0.074
VanGogh-3	7.634	0.957	0.159	0.057

5.4.2   INTERPRETING BENSE'S CHANNEL

The creative process described by Bense [16] can be further understood as the realization of an information channel between the palette and the set of regions of the image [141]. From this channel, an algorithm, which progressively partitions the image extracting all its information and revealing its structure, can be used. The rate of the information extraction will depend on the degree of order in the painting. For instance, if a painting has been created by distributing at random the colors on the canvas, any possible partition will obtain a small gain of information. However, if the painting shows a certain degree of structure, we will probably find a partition that will provide us with a larger gain of information.

The information channel $C \rightarrow R$ is defined between the random variables C (input) and R (output), which represent, respectively, the set of bins ($\mathcal{C}$) of the color histogram and the set of regions ($\mathcal{R}$) of this image. Given an image $\mathcal{I}$ of N pixels, where N_c is the frequency of bin

c ($N = \sum_{c \in C} N_c$) and N_r is the number of pixels of region r ($N = \sum_{r \in R} N_r$), the three basic elements of this channel are the following.

– The conditional probability matrix $p(R|C)$, which represents the transition probabilities from each bin of the histogram to the different regions of the image, is defined by $p(r|c) = \frac{N_{c,r}}{N_c}$, where $N_{c,r}$ is the frequency of bin c into the region r. Conditional probabilities fulfill $\forall c \in C$. $\sum_{r \in R} p(r|c) = 1$.

– The input distribution $p(C)$, which represents the probability of selecting each intensity bin c, is defined by $p(c) = \frac{N_c}{N}$.

– The output distribution $p(R)$, which represents the normalized area of each region r, is given by $p(r) = \frac{N_r}{N} = \sum_{c \in C} p(c)p(r|c)$.

The *mutual information* between C and R is defined by

$$I(C, R) = \sum_{c \in C} \sum_{r \in R} p(c,r) \log \frac{p(c,r)}{p(c)p(r)} \tag{5.8}$$

and represents the *shared information* or *correlation* between C and R.

For a decomposition of image $\mathcal{I}$ in n regions, the *ratio of mutual information* is defined by

$$M_s(n) = \frac{I(C, R)}{H(C)}, \tag{5.9}$$

where $H(C)$ is the maximum value achievable for $I(C, R)$ (when each region coincides with a pixel) [141]. The inverse function

$$M_s^{-1}\left(\frac{I(C, R)}{H(C)}\right) = n \tag{5.10}$$

gives us the number of regions obtained from a given mutual information ratio and can be interpreted as a measure of *image complexity*.

A greedy mutual-information-based algorithm [139] which splits the image in quasi-homogeneous regions is applied. This procedure takes the full image as the unique initial partition and progressively subdivides it (e.g., in a Binary Space Partition (BSP) or Quad-Tree) according to the *maximum mutual information* gain for each partitioning step. The algorithm generates a partitioning tree $T(\mathcal{I})$ for a given ratio of mutual information gain or a predefined number of regions (N_r is the number of tree leaves). This process can also be visualized from

$$H(C) = I(C, \widehat{R}) + H(C|\widehat{R}), \tag{5.11}$$

where $\widehat{R}$ is the random variable which represents the set of regions of the image and varies after each new partition. The acquisition of information increases $I(C, \widehat{R})$ (data processing inequality [42]) and decreases $H(C|\widehat{R})$, producing a reduction of uncertainty due to the equalization of the regions. Observe that the maximum mutual information that can be achieved is $H(C)$.

We consider that the resulting tree captures the structure and hierarchy of the image, and the mutual information gained in this decomposition process quantifies the capacity of an image to be ordered or the feasibility of decomposing it by an observer. Thus, the partition algorithm provides us with a way of studying the information contained in the composition of the image. The further down the regions we have to go to achieve a given level of information, the more complex the image.

Similarly to Bense's communication channel between the repertoire and the final product, the channel introduced above can be seen as the information (or communication) channel that expresses the *distribution* of colors on a canvas. Hence, given an initial entropy or uncertainty of the image and a predefined level of resolution n, the evolution of the ratio Equation 5.9 represents the distribution process. Note that n ranges from 1 to N_r so that $1 \leq n \leq N_r^{min} \leq N_r$, where N_r^{min} is the minimum number of regions that provide all the image information (i.e., $M_s(N_r^{min}) = 1$), and $N_r^{min} = N_r$ when the image has no adjacent pixels with the same color. According to the design of the channel commented on previously, the measure also works in the reverse direction. Thus, for any ratio, the corresponding $T(\mathcal{I})$ structure can be obtained where the number of tree leaves is n.

Figure 5.3 shows the evolution of M_s using a BSP partitioning procedure for each test painting. This algorithm shows us how the composition of the image (macro-aesthetic description) appears clearly after a relative small number of partitions. On the contrary, the details or forms in the painting appear when a very refined mesh is reached (micro-aesthetic description).

Observe that the capacity of extracting order from each painting coincides with the behavior expected by an observer. Note the grouping of the three different painting styles. In Table 5.3, M_s values for $n = 16$ are shown for the set of paintings. Figure 5.4 shows two resulting partitions from two van Gogh's paintings. Finally, in Figure 5.5, we show different decompositions of VanGogh-1 obtained for several values of n, and only taking into account the luminance. Each region has been painted with the average color corresponding to that region. Observe how, with a relative small number of regions, the composition of the painting is already visible (see Figures 5.5 *(c-d)*) although the details are not sufficiently represented.

The three compositional measures presented in this section capture the spatial order in an image from an informational point of view. The first two measures, M_j and M_k, measure similarities between predefined regions using the information content (Shannon entropy) and the algorithmic complexity (Kolmogorov complexity), respectively. The third measure, M_s, based on mutual information, goes one step further by dynamically evolving as the structure is being discovered.

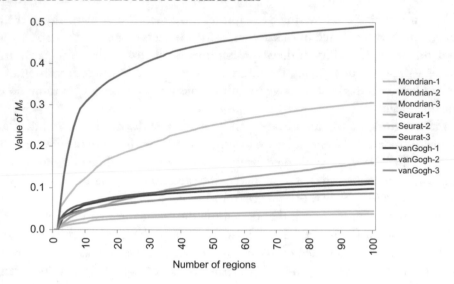

Figure 5.3: Evolution of ratio M_s for the set of test paintings. The number of regions and the value of M_s are shown on the x-axis and y-axis, respectively. The plots of Seurat-1 and Seurat-2 are overlapped ([141] © IEEE, 2008).

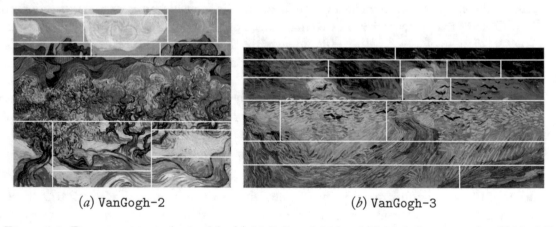

(a) VanGogh-2   (b) VanGogh-3

Figure 5.4: Decompositions obtained for (a) $M_s(16) = 0.074$ and (b) $M_s(16) = 0.057$ (see Table 5.3) ([141] © IEEE, 2008).

5.5   INFORMATIONAL ANALYSIS OF VAN GOGH'S PERIODS

The work of van Gogh has been studied extensively [22, 101, 115]. We use here a quantitative new approach based on aesthetic measures [142] to analyze the consistency of the results obtained with respect to the literature.

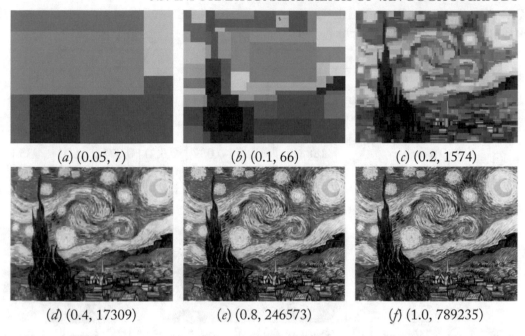

(a) (0.05, 7)   *(b)* (0.1, 66)   *(c)* (0.2, 1574)

(d) (0.4, 17309)   *(e)* (0.8, 246573)   *(f)* (1.0, 789235)

Figure 5.5: Evolution of an adaptive decomposition of VanGogh-1 using our partitioning algorithm. For each subfigure, the corresponding $(M_s(n), n)$ are shown. A total of 789,235 regions are needed to achieve the total information (i.e., N_r^{min} =789,235) ([141] © IEEE, 2008).

To study the evolution of van Gogh's style, we apply the informational aesthetic measures of Sections 5.3 and 5.4 to a subset of paintings obtained from the website *Vincent van Gogh Gallery* of David Brooks [22], where the paintings are classified chronologically in six periods (see Table 5.4). The test set has been selected keeping approximately the same proportion for each period, discarding repetitions of compositions and covering all the categories: peasants, Japonaseries, portraits, landscapes, and still lifes. This set contains 219 paintings, representing a quarter of the total number of van Gogh's paintings.

In Table 5.4 we show the average value and standard deviation for the informational measures M_B Equation 5.2, M_K Equation 5.3, and M_s^{-1} Equation 5.10 of the test set for each period. For practical purposes, we compute M_s^{-1} using luminance (X_ℓ). This is not a significant drawback as the eye is more sensitive to changes in luminance than in color and most of the information in a scene is contained in its luminance. In Figure 5.6, we show a representative painting of each period according to the period-average of the aesthetic measures (Table 5.4). Next, we analyze van Gogh's periods using the informational aesthetic measures.

Stylistically, although his genius starts to appear in the first period, some works seem flat and the colors are not always used to their best effect [22] (see Figures 5.6 *a* and 5.12 *a*). Together with period 2 (Nuenen), his palette is dark and dull in tones (Figure 5.6 *b*) but the works of period 2

Table 5.4: Average and standard deviation of the values of the informational aesthetic measures (M_B, M_K, and M_s^{-1}) for each van Gogh's period

Period			M_B		M_K		$M_s^{-1}(0.25)$	
Order	Name	Years	$\overline{x}$	$s(x)$	$\overline{x}$	$s(x)$	$\overline{x}$	$s(x)$
1	Earliest Paintings	1881-3	0.421	0.064	0.771	0.056	1072	984
2	Nuenen/Antwerp	1883-6	0.450	0.071	0.772	0.061	981	673
3	Paris	1886-8	0.378	0.061	0.704	0.073	1816	1005
4	Arles	1888-9	0.345	0.037	0.680	0.071	1823	935
5	Saint-Rémy	1889-90	0.338	0.026	0.690	0.048	2216	737
6	Auvers-sur-Oise	1890	0.324	0.033	0.677	0.072	2102	651

(*a*) Fisherman's Wife on the Beach, 1882
(0.418, 0.759, 1264)

(*b*) Shepherd with a Flock of Sheep, 1884
(0.463, 0.739, 875)

(*c*) The Seine with the Pont de la Grande Jette, 1887
(0.385, 0.718, 1396)

(*d*) Sunset: Wheat Fields Near Arles, 1888
(0.345, 0.697, 1648)

(*e*) Olive Grove: Pale Blue Sky, 1889
(0.339, 0.593, 2456)

(*f*) Daubigny's Garden, 1890
(0.315, 0.714, 2375)

Figure 5.6: A representative painting of each period is shown according to the average values of Table 5.4 (from period 1:(*a*) to 6:(*f*), © 1996–2010 David Brooks [22]). The (M_B, M_K, $M_s^{-1}(0.25)$) values are indicated for each painting.

reflect an important technical improvement as seen in his first great painting (see Figure 5.12 *b*). The style of this epoch is captured by high values of M_B, corresponding to a limited palette, high values of M_K, meaning high degree of order and easy compression (few colors and tones, and simple compositions), and low values of M_s^{-1}, reflecting a basic compositional structure (see Table 5.4 and Figure 5.7 *a*).

(*a.i*) Potato Planting, 1884   (*b.i*) Irises, 1889

(*a.ii*) $M_s^{-1}(0.25) = 29$ regions   (*b.ii*) $M_s^{-1}(0.25) = 3,378$ regions

Figure 5.7: (*i*) Two van Gogh paintings corresponding to (*a*) period 2 and (*b*) period 5, © 1996–2010 David Brooks [22]. (*ii*) Binary-space partitions for a $M_s = 0.25$ ([142] © Eurographics, 2008).

In Paris (period 3), van Gogh was influenced by Impressionism and Neo-impressionism, and his style underwent an important metamorphosis visualized by changes in the palette (from dark-hued to bright and vibrant colors), brushstroke (broken, broad, vigorous, and swirling), and subject (from peasants to Paris atmosphere). These changes also show his ongoing exploration of complementary color contrasts and a bolder style. Van Gogh wrote "I use color more arbitrarily so as to express myself more forcibly." As an example, the influence of Seurat's pointillism can be seen in Figures 5.6 *c* and 5.12 *c*. The characteristics of this period are reflected in the values of the measures of Table 5.4 with a notable jump (decreasing M_B and M_K, and increasing M_s^{-1}) because of a richer palette and a more complex composition (more details, elements, and colors).

In Provence (period 4), van Gogh progressively improves his technique and uses characteristic and intense saturated colors (Figures 5.6 *d* and 5.12 d). The work of this period reflects a synthesis of the two previous ones: Neuen and Paris. The aesthetic measures M_B and M_K decrease while M_s^{-1} increases slightly, following the tendency of the previous period.

In the short period of Saint-Rémy (period 5), van Gogh produces nice landscapes (Figures 5.6 *e* and 5.12 *e*) characterizing his style by swirls. Impressionist artists sometimes use luminance to generate the sensation of motion and van Gogh used this in a more complex way. Van Gogh's ability to depict turbulence could be due to periods of prolonged psychotic agitation. Their patterns closely follow a Kolmogorov's statistical model of turbulence obtaining a high realism [6]. With respect to the behavior of the aesthetic measures, while M_B and M_K maintain similar values to the ones of the previous period, M_s^{-1} increases due to a higher compositional complexity. This fact is illustrated by the M_s^{-1} average for this period in contrast with the same measure for the previous periods (see Table 5.4 and Figure 5.7 b).

In the final period (Auvers-sur-Oise), van Gogh could be moving into another new style [22]. He is far from the style of the initial periods (Figures 5.6 f and 5.12 f). The measures M_B and M_K achieve the lowest values reflecting their maximum distance with respect to the initial periods. On the other hand, M_s^{-1} continues reflecting a high compositional complexity.

It is interesting to observe the variation in the standard deviation of the aesthetic measures through the different periods. A higher deviation could mean more room for experimenting with palette and composition, while a lower variance would imply that the style is more defined. This matches the decrease in deviation when passing from period 3 (Paris) to period 4 (Arles). Observe also that in the M_s^{-1} measure (compositional complexity) the deviation jumps from the second to third period. This would agree with the fact that the artist would have already had a well-defined composition style in the second period, but this would have been abandoned in the experimental Paris period. Lastly, let us remark that the increase in deviation in the last period in the palette measure M_B and the palette-compositional measure M_K would match with the hypothesis by some art critics that van Gogh could have been changing his style in this last period.

Figure 5.8 shows the sequence of all selected paintings in ascending order according to the values of aesthetic measures. Each painting is depicted by a color bar which represents its period (1:yellow, 2:orange, 3:red, 4:green, 5:blue, 6:violet). In the M_B plot (Figure 5.8 a), we can observe how yellow and orange colors (initial periods) mainly correspond to the highest values in comparison to the blue and violet colors (final periods) which tend to get the lowest values. The red color represents the transition period expressed by a range of middle values mixed with the other periods. Similar tendencies are shown for the M_K measure (Figure 5.8 b). On the other hand, the compositional complexity expressed by M_s^{-1}, and calculated from three different mutual information ratios (0.15, 0.20, and 0.25), also shows a similar grouping. Thus, we can see how the lowest complexity corresponds mainly to the paintings of periods 1 and 2 and the highest complexity to the final periods. Observe that, for the plot of $M_s^{-1}(0.15)$ corresponding to a low level of captured information (Figure 5.8 c), the differences in the periods appear clearer than in the other plots with more captured information (Figures 5.8 d and 5.8 e). This is due to the fact that few partitions allow the capture of simpler composition in the first of van Gogh's periods, while the final periods need more partitions.

From the results obtained using the aesthetic measures we can conclude that the six van Gogh periods could be further grouped in three: origins (periods 1-2), transition (period 3), and maturity (periods 4-6). The transition period (Paris), where van Gogh pursued art studies and met Impressionist painters, represents a break with his previous style and a changeover to new styles. We have seen how these three periods manifest themselves in the values of our measures, which are thus able to characterize the artistic evolution of the painter.

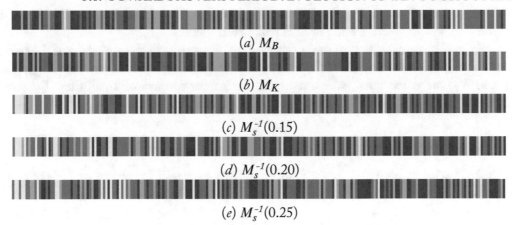

Figure 5.8: Plots of the sequence of test paintings in ascending order according to the values of aesthetic measures. Each painting is depicted by a color bar representing its period (1:yellow, 2:orange, 3:red, 4:green, 5:blue, 6:violet). M_s^{-1} has been computed for a 15, 20, and 25% of mutual information gain ([142] © Eurographics, 2008).

5.6   TOWARDS AUVERS PERIOD: EVOLUTION OF VAN GOGH'S STYLE

In the previous sections, we have presented a set of information-theoretic measures to study some informational aspects of a painting related to its palette and composition. Some of these measures, based on the entropy of the palette, the compressibility of the image, and an information channel to capture the composition of a painting, were used to discriminate different painting styles [141] and to analyze the evolution of van Gogh's artwork [142], revealing a significant correlation between the values of the measures and van Gogh's artistic periods. These measures cannot only help to categorize art into different periods [182], but also how they are able to model loci of interest when observers view an artwork, that is, where gaze is attracted in an artwork [181].

To further study the evolution of van Gogh's artwork, we use five measures based on palette entropy, compressibility, compositional complexity, randomness (entropy rate), and structural complexity (excess entropy). The first three measures have been presented in Sections 5.3 and 5.4. While the entropy of the palette only takes into account the color diversity, the other measures also consider its spatial distribution. In fact, these measures are not fully independent but offer complementary views of complexity in an image, as we will see in the analysis of the results in Section 5.6.3.

From a given color image $\mathcal{I}$ of N pixels, we use now its sRGB (see Section 5.3) and HSV representations to study the behavior of the proposed measures. HSV (hue, saturation, value) is a cylindrical-coordinate representation of sRGB which is more perceptually plausible than the sRGB cartesian representation. In this case, the alphabets are represented by $\mathcal{X}_H$, $\mathcal{X}_S$, and $\mathcal{X}_V$, according to a given discretization of each parameter.

From the normalization of the corresponding histograms of the alphabets of the color representations, the probability distributions of the corresponding random variables (X_{rgb}, X_ℓ, X_H, X_S, and X_V) are determined, which represent the *palette* features of a painting. The palette is considered as the finite and discrete range of colors used by the artist.

5.6.1   RANDOMNESS

As we have seen in Section 1.5, the joint entropy of length-L sequences or *L-block entropy* is defined by

$$H(X^L) = - \sum_{x^L \in \mathcal{X}^L} p(x^L) \log p(x^L), \tag{5.12}$$

where the sum runs over all possible L-blocks. The *entropy rate* is defined by

$$h^x = \lim_{L \to \infty} \frac{H(X^L)}{L} = \lim_{L \to \infty} h^x(L), \tag{5.13}$$

where $h^x(L) = H(X_L | X_{L-1}, X_{L-2}, \ldots, X_1)$ is the entropy of a symbol conditioned on a block of $L - 1$ adjacent symbols. The entropy rate of a sequence measures the average amount of information (i.e., irreducible randomness) per symbol x and the optimal achievement for any possible compression algorithm [42, 53]. Entropy rate can be also seen as the uncertainty associated with a given symbol if all the preceding symbols are known. For more details, see Section 1.5.

The entropy rate of an image quantifies the average uncertainty surrounding a pixel, that is, the difficulty of predicting the color of its neighbor pixels. While a painting that is highly random is difficult to compress, a painting with low randomness has many correlations with pixel colors. It is interesting to note that $\log |\mathcal{C}| - h^x$ can be also considered as a measure of redundancy in a painting.

In the context of an image, $\mathcal{X}$ represents the color alphabet and x^L is given by a set of L *neighbor* pixel intensity values. In practice, we cannot compute L-block entropies for high L, due to the exponential size $-N^L$, where N is the cardinality of $\mathcal{X}$— of the joint histogram. In our tests (see Sec. 5.6.3), the entropy rate has been estimated taking L-block samples radially around each pixel. This pixel represents the origin and becomes the first element of the block. To carry out the computations, we set $L = 3$ and $N = 256$. Using digital photography software, we have conducted experiments that showed a positive correlation between entropy rate and contrast.

5.6.2 STRUCTURAL COMPLEXITY

A complementary measure to the entropy rate is the *excess entropy*, which is a measure of the *structure* of a system (see Section 1.5). The *excess entropy* is defined by

$$E = \sum_{L=1}^{\infty} (h^x(L) - h^x) \tag{5.14}$$

$$= \lim_{L \to \infty} (H(X^L) - h^x L) \tag{5.15}$$

and captures how $h^x(L)$ converges to its asymptotic value h^x.

Considered by many authors as a measure of the structural complexity of a system, the excess entropy is introduced here to measure the spatial structure of a painting. If excess entropy is large the painting contains many regularities or correlations [54]. Thus, excess entropy serves to detect ordered, low entropy density patterns in a painting. In the case of a completely random image, the excess entropy should vanish, showing that correlations are not present in the image. In our tests, the excess entropy has been estimated using Equation 5.15 and taking $L = 5$ and $N = 32$. While $L = 3$ is enough to compute the entropy rate, excess entropy needs larger sequences, which implies reducing the number of bins due to computational restrictions.

5.6.3 ARTISTIC ANALYSIS

In this section, we analyze how the style of van Gogh evolves toward his last period. According to art critics, in Auvers, van Gogh changes his style in the following way: he sees the Northern landscape with a sharpened and heightened vision; softens the hue in landscapes (reflecting the response to the more subdued Northern light with whites, blues, violets, and soft greens); uses harsher primary colors; exhibits a certain unevenness and impetuosity of brushstroke; and simplifies the composition (see Ronald Pickvance [127]). Can our measures support these claims on the evolution of the palette and composition?

To analyze the evolution of van Gogh's style, a set of measures presented in this chapter have been applied to the images of van Gogh's paintings obtained from *The Vincent van Gogh Gallery* of David Brooks [22]. In this website, van Gogh's oeuvre (861 paintings) is classified into six periods. From this set of images, we have excluded 61 black and white images which were not available in color yielding a total of 800 color images for our experiments.

We will first consider the palette measures. From the entropy-based measure M_B (Table 5.5), we can see how the palette evolves. There is a first palette simplification from the Early to Nuenen period, but starting with the Paris-period the palette entropy constantly decreases, obtaining its minimum in Auvers. It is important to note that the measure of entropy is logarithmic, that is, the constant although small increases in the measure translates into a much larger absolute increase in the variety of colors used.

In Figure 5.9 we show the digital-image-palette (DIP, see Appendix A) of each period based on the HSV representation. In this figure, we show a painting of each period (a), the DIP

Table 5.5: For each period and the global artwork, number of paintings, canvas size average (dm²), M_B, M_K, and M_s^{-1}(0.1, 0.15, 0.20, and 0.25 are shown ($N = 256$ bins has been used)

Period	#	Size	M_B	M_K	$M_s^{-1}(0.05)$	$M_s^{-1}(0.1)$	$M_s^{-1}(0.15)$	$M_s^{-1}(0.20)$	$M_s^{-1}(0.25)$
Early	26	17.5	0.422	0.769	6.154	39.769	147.538	413.538	1019.000
Nuenen	172	23.7	0.486	0.794	5.727	33.878	153.953	479.378	1144.616
Paris	209	22.0	0.384	0.712	12.177	81.301	309.541	813.287	1688.938
Arles	181	40.2	0.351	0.688	13.376	93.834	344.094	869.221	1748.387
St-Rémy	137	42.8	0.342	0.665	27.766	185.985	587.219	1286.766	2331.175
Auvers	75	39.5	0.334	0.659	27.613	164.307	509.187	1130.560	2081.400
Global	800	31.1	0.388	0.713	14.983	98.300	344.911	851.988	1710.363

of this painting (b), the DIP of the period (c), and the normalized DIP (NDIP) of the period. The DIP representation has been obtained from a discretization of the hue in 360 bins ($\mathcal{X}_H$) and, for each bin, the average of both saturation and brightness is depicted together with the hue. The average of the achromatic values is represented by the gray-color of the circumference. For each painting, the frequency of bins has been weighted by the real size of canvas. Observe that the canvas size has been doubled from Paris on (Table 5.5). In the last row of Figure 5.9, the global palette of all periods is shown. As the figure shows, the palette gains in chromaticity (except for the somber palette of the Nuenen period) and evolves toward softer colors, becoming more and more constrained in hue space. At the same time, the palette also evolves toward more yellowish and brighter hues overall. All of this means that van Gogh was continuously evolving and optimizing his palette. Also, let us note the remarkable similarity of the global average to the Paris one, especially striking in the NDIP (Figure 5.9 d) —in a way the Paris period represents van Gogh's oeuvre remarkably well.

To quantify the palette difference between periods we use a DIP-distance defined in Equation A.1. In Table 5.6, we show the distances between the DIPs of all periods and global artwork. The Nuenen period has the maximum average distance to the other periods, while the Paris period yields the minimum distance, even to the global palette, reinforcing the central role of Paris period in van Gogh's artwork. Interestingly, the distances from the Auvers period are more balanced toward all the other periods, being of course closer to Saint-Rémy which could be due to the fact that in Auvers van Gogh reflected on all of his previous periods. Indeed, before going to Auvers, he spent some days in Paris and had the opportunity of reviewing a large part of his previous paintings, as he explains in a letter to his sister Wil [162].

With respect to composition, we can group the six van Gogh periods into three distinct groups (see Table 5.5): Early/Nuenen, Paris/Arles, and Saint-Rémy/Auvers. The composition from one group to the other one exhibits large changes, doubling (for low mutual information ratios as 0.05 and 0.1) the number of regions to extract the same amount of information. We see a peak of compositional complexity in Saint-Rémy period, followed by a slight decrease in Auvers. Again, this quantitative findings is in accordance with critics opinion about this period with respect to the simplification of composition [127].

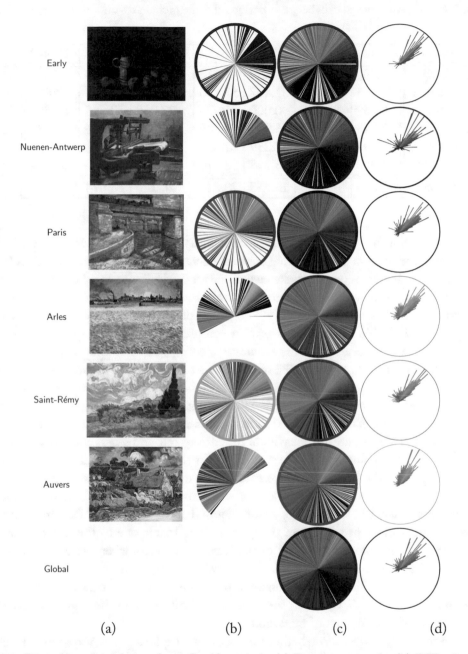

Figure 5.9: Digital-image-palette of van Gogh's periods. (a) Painting example. (b) DIP of painting (a). (c) DIP of the period. (d) NDIP of the period. Global DIP and NDIP are shown in the last row. Painting images credit: © 1996–2010 David Brooks [22]([143] © Eurographics, 2010).

Table 5.6: DIP-distance matrix between periods and the global artwork. The average column is only computed from period columns

Period	Early	Nuenen	Paris	Arles	St-Rémy	Auvers	Global	Avg
Early	0.000	26.775	28.559	47.213	37.197	44.498	25.658	36.848
Nuenen	26.775	0.000	31.437	57.830	52.633	53.289	29.425	44.393
Paris	28.559	31.437	0.000	35.976	37.259	43.895	13.955	35.425
Arles	47.213	57.830	35.976	0.000	23.857	38.076	29.824	40.590
St-Rémy	37.197	52.633	37.259	23.857	0.000	31.401	28.120	36.469
Auvers	44.498	53.289	43.895	38.076	31.401	0.000	36.074	42.232
Global	25.658	29.425	13.955	29.824	28.120	36.074	0.000	32.611

If we accept that entropy rate measure positively correlates with contrast (see Section 5.7), then we can obtain from Table 5.7 that contrast decreases from the Early period to Nuenen but later constantly increases (entropy rate evolves inversely similar to the palette redundancy M_B). The entropy rates achieve their maximum values in the last period, which is again in accordance to art critics' prevalent analysis: the simplification of composition was accompanied by an increase in contrast [127]. Table 5.5 also shows how the complexity M_K, which expresses the compression ratio, behaves in an inverse way to the evolution of entropy rate h^x. This behavior agrees with the fact that the entropy rate expresses the optimal achievement for a compression algorithm.

As we interpret excess entropy as a measure of the degree of correlation and patterns, we can read from Table 5.7 how the Auvers period presents more brightness patterns, while Arles period shows more structure in chromaticity (hue and saturation). In Figure 5.10 we show two paintings of Auvers period to illustrate the behavior of the entropy rate and excess entropy. Observe first that entropy rates of top painting are higher, specially for the brightness. This matches with the high contrasted spots in the foliage of the trees due to the diversity in the illumination and chroma of the leaves. On the other hand, the sheaves of wheat and the background present a more uniform color which translates in lower entropy rate values. The excess entropy of the top painting is also higher revealing more patterns than the bottom one. This is due to the fact that the apparent randomness of the color of the pixel of the leaves disappears when we take into account the correlations in the sequences of pixels. This is, we discover, order out of apparent randomness. In the bottom image, either the sequences of pixels studied are too short, due to computational limitations ($L = 5$ and $N = 32$), or the uniformity is higher from the beginning. In either case, the uniformity discovered out of randomness is lower.

Addressing the question whether van Gogh was exploring new ways toward changing his style, we can answer for the Auvers period that the measures, indeed, reflect the fact that van Gogh traded off simplified composition against an extended palette and increased contrast. Furthermore, palette extension, contrast increase, and compositional complexity increase can be seen as van Gogh's aesthetic development from his Paris period to Saint-Rémy.

Table 5.7: For each period and the global artwork, average of entropy rate h^x and excess entropy E for hue (H), saturation (S), and brightness value (V) are shown. Entropy rate has been computed using $L = 3$ and $N = 256$, and excess entropy using $L = 5$ and $N = 32$. The standard deviation is shown for each measure

| Period | h^x_H | | h^x_S | | h^x_V | | E_H | | E_S | | E_V | |
Name	$\overline{x}$	$s(x)$	$\overline{x}$	$s(x)$	$\overline{x}$	$s(x)$	$\overline{x}$	$s(x)$	$\overline{x}$	$s(x)$	$\overline{x}$	$s(x)$
Early	5.251	0.724	6.551	0.443	6.802	0.452	1.791	0.376	1.785	0.523	1.961	0.477
Nuenen	4.899	0.810	6.219	0.602	6.215	0.776	1.739	0.345	1.590	0.557	1.669	0.534
Paris	5.564	0.709	6.802	0.469	6.972	0.405	1.771	0.360	1.898	0.477	1.886	0.382
Arles	5.855	0.598	7.034	0.305	7.272	0.286	1.935	0.309	2.063	0.381	2.151	0.366
St-Rémy	5.859	0.652	6.972	0.322	7.427	0.248	1.829	0.354	1.906	0.385	2.180	0.312
Auvers	5.925	0.531	7.116	0.309	7.471	0.305	1.880	0.279	2.009	0.339	2.223	0.305
Global	5.561	0.804	6.780	0.502	6.996	0.547	1.822	0.361	1.877	0.492	1.984	0.520

Figure 5.10: Entropy rate and excess entropy values of two paintings of Auvers period: (top) $h^x_{HSV} = (5.718, 7.696, 9.279)$ and $E_{HSV} = (1.189, 3.602, 5.719)$; (bottom) $h^x_{HSV} = (4.156, 7.215, 5.000)$ and $E_{HSV} = (0.492, 1.733, 0.366)$.

As we have seen, the case of the Paris period is interesting in that, for almost all considered measures, this period closely approximates the global average. Given that for art critics this period constitutes an exploratory phase for van Gogh, our results show that in this period, indeed, past and future styles are being contained and tested.

5.7   COLOR AND REGIONAL INFORMATION

In this section, we study how the information is distributed on the painting by computing the information associated with each color and region [142].

We now focus our attention on the mutual information between C and R, that expresses the degree of *dependence* or *correlation* between the set of color bins and the regions of the painting. From Equation 5.8, mutual information can be expressed as

$$
\begin{aligned}
I(C; R) &= \sum_{c \in C} p(c) \sum_{r \in R} p(r|c) \log \frac{p(r|c)}{p(r)} \\
&= \sum_{c \in C} p(c) I(c, R),
\end{aligned}
\tag{5.16}
$$

where we define

$$
I(c; R) = \sum_{r \in R} p(r|c) \log \frac{p(r|c)}{p(r)}
\tag{5.17}
$$

as the *color mutual information* (CMI), which gives us the degree of dependence between the color c and the regions of the painting, and is interpreted as a measure of the *information or saliency* associated with color c.

As we have noted in Section 1.3, $I(c; R)$ can be expressed as a Kullback-Leibler distance (Section 1.2). Thus, $I(c; R) = KL(p(R|c)|p(R))$, where $p(R|c)$ (true p.d.) is the conditional probability distribution between c and the painting regions, and $p(R)$ is the marginal probability distribution of R, which in our case corresponds to the distribution of region areas (target p.d.). According to this, high values of CMI express a high dependence or correlation between a color and a given region, and identify the most relevant colors, that is, colors conveying more information. On the other hand, the lowest values correspond to the colors distributed uniformly in the painting.

Similarly, the information associated with a region can be defined from the inverted channel $R \rightarrow C$, so that R is the input and C the output. From the Bayes' theorem, $p(c, r) = p(c)p(r|c) = p(r)p(c|r)$, the mutual information (Equation 5.16) can be rewritten as

$$
\begin{aligned}
I(R; C) &= \sum_{r \in R} p(r) \sum_{c \in C} p(c|r) \log \frac{p(c|r)}{p(c)} \\
&= \sum_{r \in R} p(r) I(r, C),
\end{aligned}
\tag{5.18}
$$

where we define

$$
I(r; C) = \sum_{c \in C} p(c|r) \log \frac{p(c|r)}{p(c)}
\tag{5.19}
$$

as the *regional mutual information* (RMI), which represents the degree of correlation between the region r and the set of color bins, and can be interpreted as the information or saliency associated with region r. Analogous to CMI, low values of RMI correspond to regions that have an approximated representation of the palette (i.e., $p(C|r)$ is close to $p(C)$). On the other hand, high values correspond to regions that have few and exclusive colors.

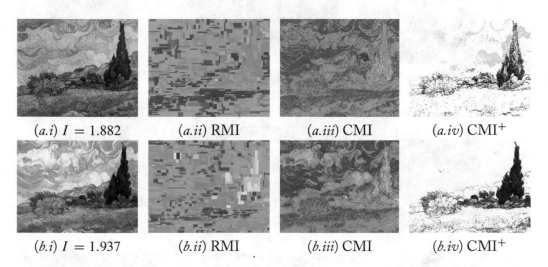

| *(a.i)* $I = 1.882$ | *(a.ii)* RMI | *(a.iii)* CMI | *(a.iv)* CMI$^+$ |
| *(b.i)* $I = 1.937$ | *(b.ii)* RMI | *(b.iii)* CMI | *(b.iv)* CMI$^+$ |

Figure 5.11: (*i*) Two van Gogh's paintings of period 4: (*a.i*) Wheat Field with Cypresses at the Haute Galline Near Eygalieres, June, 1889, and (*b.i*) Wheat Field with Cypresses, September, 1889, © 1996–2010 David Brooks [22]. (*ii*) Regional and (*iii*) color mutual information maps calculated using a 25% of mutual information gain. (*iv*) From the original painting, only the most salient pixels (its CMI is in the upper half range) have been depicted ([143] © Eurographics, 2010).

In Figure 5.11 *a*, we show a composition with a cypress which was an element associated with death that obsessed van Gogh. This work was painted in June 1889, after van Gogh's arrival at Saint-Rémy. Months before, he painted two variants of this composition. One of these is presented in Figure 5.11 *b*. For both paintings shown, the information associated with each painting region (RMI) and with each color (CMI) is indicated. In the first case (Figure 5.11 *ii*), a thermic scale (from blue to red) is used to represent the RMI values. In the second case (Figure 5.11 *iii*), each pixel of the painting is visualized with the CMI value associated with its luminance. In addition, the images in Figure 5.11 *iv* have been created depicting only the original pixels with CMI values in the upper half part of the CMI range (CMI$^+$). That is, only the most salient pixels of the painting are shown. Observe that, in spite of being the same composition, it is easy to see the informational differences between both paintings. For instance, the cypress is more salient in Figure 5.11 *b* than in Figure 5.11 *a* while some clouds are more salient in Figure 5.11 *a*.

In Figure 5.12, we show the CMI maps for a painting of each period representing different categories. Following the sequence of paintings, some of the most salient elements are the path and faces (Figure 5.12 *a*), the lamp and most illuminated areas (Figure 5.12 *b*), the eyes, mouth,

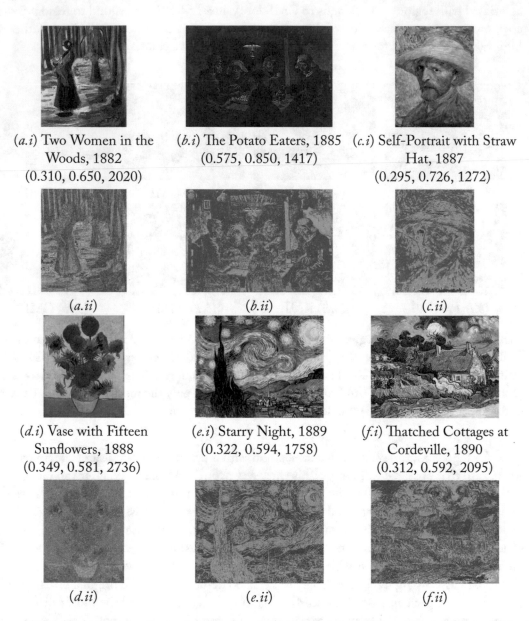

(*a.i*) Two Women in the
Woods, 1882
(0.310, 0.650, 2020)

(*b.i*) The Potato Eaters, 1885
(0.575, 0.850, 1417)

(*c.i*) Self-Portrait with Straw
Hat, 1887
(0.295, 0.726, 1272)

(*a.ii*)

(*b.ii*)

(*c.ii*)

(*d.i*) Vase with Fifteen
Sunflowers, 1888
(0.349, 0.581, 2736)

(*e.i*) Starry Night, 1889
(0.322, 0.594, 1758)

(*f.i*) Thatched Cottages at
Cordeville, 1890
(0.312, 0.592, 2095)

(*d.ii*)

(*e.ii*)

(*f.ii*)

Figure 5.12: (*i*) An outstanding painting of each period is shown (from period 1:(*a*) to 6:(*f*), ©
1996–2010 David Brooks [22]). The $(M_B, M_K, M_S^{-1}(0.25))$ values are indicated for each painting. (*ii*)
Visualization of the mutual information associated with each color (CMI) using a thermic scale for
each painting ([143] © Eurographics, 2010).

and hat (Figure 5.12 *c*), the petals of the sunflowers (Figure 5.12 *d*), the moonlight and cypress (Figure 5.12 *e*), and the cloud, soil, and house edges (Figure 5.12 *f*).

APPENDIX A

Digital-Image-Palette

In order to represent the palette of an image, we define the Digital-Image-Palette (DIP) based on the following rules.

- The HSV color representation is selected to depict the colors of the palette with hue, saturation, and value. We consider the cylindric representation with $H \times S \times V$ in the range $[0°, 360°) \times [0, 1] \times [0, 1]$.

- The hue h of an hsv value refers to a pure color without tint or shade (addition of white or black pigment, respectively); the value v represents the brightness relative to the brightness of a similarly illuminated white; and the saturation s represents the colorfulness relative to its own brightness v.

- The space is discretized into N bins H_i (e.g., 360) where each one corresponds to a cylindrical sector. A bin H_i represents all the colors that fall inside it.

- The achromatic colors (gray-scale) have an undefined hue and a null saturation. Thus, we consider N chromatic bins and one achromatic: $M = N + 1$ bins.

- F_i is the frequency of H_i weighted by the real size of canvas in order to avoid the heterogeneous scales of the images with respect to the real size of the paintings.

- The hue h_i assigned to a sector H_i is given by the angle of the middle of the arc of the sector.

- A point in the HSV space is projected into the plane $S \times V$ of its corresponding H_i. This projected point is represented by $\vec{sv}$ containing the saturation and brightness information.

  The DIP is obtained according to the next steps.

1. For each pixel $p \in \mathcal{I}$ do:

   (a) $hsv = HSV(RGB(p))$,

   (b) $H_i \leftarrow h, \ i \in \{1, \ldots, M\}$,

   (c) increase F_i, and

   (d) add $\vec{sv}$ into H_i.

2. For each H_i do:

    (a) $\overrightarrow{SV}_i$ = vectorial sum for all $\overrightarrow{sv}$ in H_i,

    (b) $\overrightarrow{sv}_i$ = normalization of $\overrightarrow{SV}_i$ from F_i,

    (c) $hsv_i = (h_i, \pi_1(\overrightarrow{sv}_i), \pi_2(\overrightarrow{sv}_i))$, and

    (d) paint sector H_i with color hsv_i.

The visual representation of a DIP is composed by the set of sectors H_i in a circle of unitary radius for the chromatic colors, and by a circumference painted with the achromatic value (Figure A.1). The frequency F_i is normalized (f_i) to represent a normalized DIP (NDIP). Its visualization uses variable radius (chromatic colors) and the circumference width (achromatic colors) to express that normalization.

The dissimilarity, or DIP-distance, between two DIPs i and j is defined by

$$d_{ij} = \frac{1}{M} \sum_{k=1}^{M} |f_{i_k} \times \overrightarrow{sv}_{i_k} - f_{j_k} \times \overrightarrow{sv}_{j_k}|. \tag{A.1}$$

Figure A.1: DIP representation examples. For all hues, (left) $s = 0.5$ and $v = 1$, (center) $s = 1$ and $v = 1$, and (right) $s = 1$ and $v = 0.5$. The achromatic value v is represented on the border of the circumference.

Bibliography

[1] The Open Video Project: a shared digitial video collection. `http://www.open-video.org/index.php`. 84, 90

[2] TREC Video Retrieval Evaluation: TRECVID. `http://trecvid.nist.gov/`. 90, 91

[3] TRECVID 2007 shot boundary determination results. `http://www-nlpir.nist.gov/projects/tv2007/active/\results/shot.boundaries/runTable.web`. 90

[4] A. S. Abutaleb. Automatic thresholding of gray-level pictures using two-dimensional entropy. *Computer Vision, Graphics, and Image Processing*, 47(1):22–32, 1989. DOI: 10.1016/0734-189X(89)90051-0. 59

[5] M. S. Ali and S. D. Silvey. A general class of coefficient of divergence of one distribution from another. *Journal of Royal Statistical Society (Serie B)*, 28(1):131–142, 1966. 19

[6] J. L. Aragón, G. G. Naumis, M. Bai, M. Torres, and P. K. Maini. Turbulent luminance in impassioned van Gogh paintings. *Journal of Mathematical Imaging and Vision*, 30(3):275–283, 2008. DOI: 10.1007/s10851-007-0055-0. 114

[7] A. Bardera, I. Boada, M. Feixas, and M. Sbert. Image segmentation using excess entropy. *Journal of Signal Processing Systems*, 54(1-3):205–214, January 2009. DOI: 10.1007/s11265-008-0194-6. 61, 62, 63

[8] A. Bardera, M. Feixas, and I. Boada. Normalized similarity measures for medical image registration. In *Proceedings of Medical Imaging SPIE 2004*, volume 5370, pages 108–118, San Diego, USA, February 2004. DOI: 10.1117/12.536106. 30, 41

[9] A. Bardera, M. Feixas, I. Boada, J. Rigau, and M. Sbert. Registration-based segmentation using the information bottleneck method. In *Iberian Conference on Patern Recognition and Image Analisys (IbPRIA 2007), Proceedings*, volume 4478-II of *Lecture Notes in Computer Science*, pages 190–197. Girona, Spain, June 2007. DOI: 10.1007/978-3-540-72849-8_17. 70

[10] A. Bardera, M. Feixas, I. Boada, and M. Sbert. Medical image registration based on random line sampling. In *IEEE International Conference on Image Processing (ICIP'05), Proceedings*, Genova, Italy, September 2005. DOI: 10.1109/ICIP.2005.1529961. 36

[11] A. Bardera, M. Feixas, I. Boada, and M. Sbert. Compression-based image registration. In *IEEE International Symposium on Information Theory*, pages 436–440, Seattle, USA, July 2006. DOI: 10.1109/ISIT.2006.261706. 42, 43

[12] A. Bardera, M. Feixas, I. Boada, and M. Sbert. High-dimensional normalized mutual information for image registration using random lines. In *Proceedings of the Third International Workshop on Biomedical Image Registration (WBIR 2006)*, volume 4057 of *Lecture Notes in Computer Science*, pages 264–271. Utrecht, Netherlands, July 2006. DOI: 10.1007/11784012_32. 38

[13] A. Bardera, M. Feixas, I. Boada, and M. Sbert. Image registration by compression, *Information Sciences*, 180(7):1121–1133, 2010. DOI: 10.1016/j.ins.2009.11.031. 42, 44

[14] A. Bardera, J. Rigau, I. Boada, M. Feixas, and M. Sbert. Image segmentation using information bottleneck method. *IEEE Transactions on Image Processing*, 2009. DOI: 10.1109/TIP.2009.2017823. 65, 67, 69, 70, 73

[15] C. Bennett, P. Gács, M. Li, P. Vitányi, and W. Zurek. Information distance. *IEEE Transactions on Information Theory*, 44(4):1407–1423, July 1998. DOI: 10.1109/18.681318. 22, 42

[16] M. Bense. *Einführung in die informationstheoretische Ästhetik. Grundlegung und Anwendung in der Texttheorie (Introduction to the Information-theoretical Aesthetics. Foundation and Application in the Text Theory)*. Rowohlt Taschenbuch Verlag GmbH., 1969. 97, 98, 107

[17] P. Bernaola, J. L. Oliver, and R. Román. Decomposition of DNA sequence complexity. *Physical Review Letters*, 83(16):3336–3339, October 1999. DOI: 10.1103/PhysRevLett.83.3336. 66

[18] M. Bezzi. Quantifying the information transmitted in a single stimulus. *Biosystems*, 89(1-3):4–9, May-June 2007. DOI: 10.1016/j.biosystems.2006.04.009. 46, 48

[19] G. D. Birkhoff. *Aesthetic Measure*. Harvard University Press, Cambridge (MA), USA, 1933. DOI: 10.4159/harvard.9780674734470. 97, 98

[20] R. Bramon, I. Boada, A. Bardera, J. Rodriguez, M. Feixas, J. Puig, and M. Sbert. Multimodal data fusion based on mutual information. *IEEE Transactions on Visualization and Computer Graphics*, 18(9):1574–1587, 2012. DOI: 10.1109/TVCG.2011.280. 45, 49, 51

[21] A. Brink. Minimum spatial entropy threshold selection. *IEE Proceedings-Vision, Image, and Signal Processing*, 142(3):128–132, June 1995. DOI: 10.1049/ip-vis:19951850. 61

[22] D. Brooks. The Vincent van Gogh Gallery. http://www.vggallery.com, 2010. xv, 97, 98, 110, 111, 112, 113, 114, 117, 119, 123, 124

[23] J. Burbea and C. R. Rao. On the convexity of some divergence measures based on entropy functions. *IEEE Transactions on Information Theory*, 28(3):489–495, May 1982. DOI: 10.1109/TIT.1982.1056497. 12, 41

[24] D. A. Butts. How much information is associated with a particular stimulus? *Network: Computation in Neural Systems*, 14:177–187, 2003. DOI: 10.1088/0954-898X/14/2/301. 8, 9, 10, 46, 47, 48

[25] T. Butz and J. Thiran. Affine registration with feature space mutual information. In *International Conference on Medical Image Computing and Computed Assisted Intervention (MICCAI 2001), Proceedings*, Lecture Notes in Computer Science, pages 549–556. SPIE, 2001. DOI: 10.1007/3-540-45468-3_66. 39

[26] T. Butz and J.-P. Thiran. Shot boundary detection with mutual information. In *International Conference on Image Processing, ICIP '01*, pages 422–425, 2001. DOI: 10.1109/ICIP.2001.958141. 76

[27] W. Cai and G. Sakas. Data intermixing and multi-volume rendering. *Computer Graphics Forum.*, 18:359–368, 1999. DOI: 10.1111/1467-8659.00356. 44

[28] F. Calderero and F. Marqués. Region merging techniques using information theory statistical measures. *IEEE Transactions on Image Processing*, 19(6):1567–1586, 2010. DOI: 10.1109/TIP.2010.2043008. 68

[29] V. Caselles, F. Catte, T. Coll, and F.Dibos. A geometric model for active contours. *Numerische Mathematik*, 66:1–31, 1993. DOI: 10.1007/BF01385685. 63

[30] F. Castro, R. Martínez, and M. Sbert. Quasi-Monte Carlo and extended first shot improvements to the multi-path method. In *Proceedings of Spring Conference on Computer Graphics'98*, pages 91–102, April 1998. Held in Budmerice, Slovak Republic. 38

[31] Z. Cernekova, I. Pitas, and C. Nikou. Information theory-based shot cut/fade detection and video summarization. *IEEE Transactions on Circuits and Systems Video Technology*, 16(1):82–91, January 2006. DOI: 10.1109/TCSVT.2005.856896. 76, 80, 83, 90, 93, 94

[32] B. Chanda and D. Majumder. A note on the use of graylevel co-occurence matrix in threshold selection. *Signal Processing*, 15(2):149–167, 1988. DOI: 10.1016/0165-1684(88)90067-9. 58

[33] C. Chang, K. Chen, J. Wang, and M. L. G. Althouse. A relative entropy-based approach to image thresholding. *Pattern Recognition*, 27(9):1275–1289, September 1994. DOI: 10.1016/0031-3203(94)90011-6. 58, 59

[34] C. I. Chang, Y. Du, J. Wang, S. M. Guo, and P. D. Thouin. Survey and comparative analysis of entropy and relative entropy thresholding techniques. *Vision, Image and Signal Processing, IEE Proceedings -*, 153(6):837–850, 2006. DOI: 10.1049/ip-vis:20050032. 54

[35] H. S. Chang, S. Sull, and S. U. Lee. Efficient video indexing scheme for content-based retrieval. *IEEE Trans Circ Syst Video Technol*, 9(8):1269–1279, December 1999. DOI: 10.1109/76.809161. 84

[36] R. Cilibrasi, R. de Wolf, and P. Vitányi. Algorithmic clustering of music based on string compression. *Computer Music Journal*, 28(4):49–67, 2004. DOI: 10.1162/0148926042728449. 42

[37] R. Cilibrasi and P. M. B. Vitányi. Clustering by compression. *IEEE Transactions on Information Theory*, 51(4):1523–1545, April 2005. DOI: 10.1109/TIT.2005.844059. 42

[38] C. Cocosco, V. Kollokian, R.-S. Kwan, and A. Evans. Brainweb: Online interface to a 3D MRI simulated brain database. *NeuroImage*, 5(4):S425, 1997. 30

[39] L. Cohen. On active contour models and balloons. *Computer Vision, Graphics and Image Processing : Image Understanding*, 5:211–218, March 1991. DOI: 10.1016/1049-9660(91)90028-N. 63

[40] A. Collignon. *Multimodality medical image registration by maximization of mutual information*. PhD thesis, Catholic University of Leuven, Leuven, Belgium, May 1998. 34

[41] A. Collignon, D. Vandermeulen, P. Suetens, and G. Marchal. 3D multi-modality medical image registration using feature space clustering. In S. Verlag, editor, *First International Conference on Computer Vision, Virtual Reality and Robotics in Medicine*, volume 905 of *Lecture Notes in Computer Science*, pages 195–204. Nice, France, April 1995. 31, 33

[42] T. M. Cover and J. A. Thomas. *Elements of Information Theory*. Wiley Series in Telecommunications, 1991. DOI: 10.1002/0471200611. 1, 3, 5, 10, 11, 12, 13, 14, 15, 16, 33, 47, 77, 97, 108, 116

[43] J. P. Crutchfield and D. P. Feldman. Regularities unseen, randomness observed: Levels of entropy convergence. *Chaos*, 15:25–54, 2003. DOI: 10.1063/1.1530990. 13, 14

[44] I. Csiszár. Eine Informationsheoretische Ungleichung und ihre Anwendungen auf den Beweis der Ergodizität von Markoffschen Ketten[1]. *Magyar Tudományos Akadémia Közleményei*, 8:85–108, 1963. 19

[45] I. Csiszár and P. C. Shields. Information theory and statistics: A tutorial. *Foundations and Trends in Communications and Information Theory*, 1(4), 2004. DOI: 10.1561/0100000004. 7, 19, 20

---

[1]An information-theoretic inequality and its application to the proof of the ergodicity of Markov chains.

[46] Z. Dawy, J. Hagenauer, P. Hanus, and J. C. Mueller. Mutual information based distance measures for classification and content recognition with applications to genetics. In *IEEE International Conference on Communications (ICC 2005)*, volume 2, pages 820–824, Seoul, Korea, May 2005. DOI: 10.1109/ICC.2005.1494466. 43

[47] M. R. Deweese and M. Meister. How to measure the information gained from one symbol. *Network: Computation in Neural Systems*, 10(4):325–340, November 1999. DOI: 10.1088/0954-898X/10/4/303. 8, 9, 10, 46, 47, 48

[48] I. S. Dhillon, S. Mallela, and D. S. Modha. Information-theoretic co-clustering. In *Proceedings of The 9th ACM SIGKDD International Conference on Knowledge Discovery and Data Mining (KDD-2003)*, pages 89–98, New York (NY), USA, 2003. ACM Press. DOI: 10.1145/956750.956764. 71

[49] V. Duay, S. Luti, G. Menegaz, and J.-P. Thiran. Active Contours and Information Theory for Supervised Segmentation on Scalar Images. In *15th European Signal Processing Conference (EUSIPCO)*, Poznan, Poland, 2007. 64

[50] D. M. Endres and J. E. Schindelin. A new metric for probability distributions. *IEEE Transactions on Information Theory*, 49(7):1858–1860, 2003. DOI: 10.1109/TIT.2003.813506. 20

[51] F. Escolano, P. Suau, and B. Bonev. *Information Theory in Computer Vision and Pattern Recognition*, Springer, 2009. 29, 53

[52] D. Feldman and J. Crutchfield. Structural information in two-dimensional patterns: Entropy convergence and excess entropy. *Physical Review E*, 67, 2003. DOI: 10.1103/PhysRevE.67.051104. 61

[53] D. P. Feldman. A brief introduction to: Information theory, excess entropy and computational mechanics. Lecture notes, Department of Physics, University of California, Berkeley (CA), USA, 2002. http://hornacek.coa.edu/dave/. 13, 14, 15, 116

[54] D. P. Feldman, C. S. McTague, and J. P. Crutchfield. The organization of intrinsic computation: Complexity-entropy diagrams and the diversity of natural information processing. *Chaos*, 18:043106, October 2008. DOI: 10.1063/1.2991106. 14, 117

[55] J. Freixenet, X. Muñoz, D. Raba, J. Martí, and X. Cufí. Yet another survey on image segmentation: Region and boundary information integration. In *European Conference on Computer Vision*, pages 408–422, Copenhagen, Denmark, May 2002. DOI: 10.1007/3-540-47977-5_27. 53

[56] R. Fuchs and H. Hauser. Visualization of multivariate scientific data. *Computer Graphics Forum*, 28(6):1670–1690, 2009. DOI: 10.1111/j.1467-8659.2009.01429.x. 44

134 BIBLIOGRAPHY

[57] S. Furuichi. Information theoretical properties of Tsallis entropies. *Journal of Mathematical Physics*, 47(2), 2006. DOI: 10.1063/1.2165744. 22

[58] R. Gan and A. Chung. Multi-dimensional mutual information based robust image registration using maximum distance-gradient-magnitude. In *19th International Conference on Information Processing in Medical Imaging 2005 (IPMI'05)*, volume 3565 of *Lecture Notes in Computer Science*, pages 210–221. July 2005. DOI: 10.1007/11505730_18. 39

[59] U. Gargi, R. Kasturi, and S. Strayer. Performance characterization of video-shot-change detection methods. *IEEE Transactions on Circuits and Systems for Video Technology*, 10(1):1–13, February 2000. DOI: 10.1109/76.825852. 77

[60] R. C. Gonzalez and R. E. Woods. *Digital Image Processing*. Prentice Hall, Upper Saddle River (NJ), USA, 2002. 53, 58, 80

[61] R. M. Gray. *Entropy and Information Theory*. Springer-Verlag, New York (NY), USA, 1990. DOI: 10.1007/978-1-4757-3982-4. 17

[62] G. Greenfield. On the origins of the term "computational aesthetics". In L. Neumann, M. Sbert, B. Gooch, and W. Purgathofer, editors, *Computational Aesthetics 2005. Eurographics Workshop on Computational Aesthetics in Graphics, Visualization and Imaging*, pages 9–12. Eurographics Association, May 2005. 99

[63] A. Hanjalic. Shot-boundary detection: unraveled and resolved? *IEEE Transactions on Circuits and Systems for Video Technology*, 12(2):90–105, 2002. DOI: 10.1109/76.988656. 75

[64] R. M. Haralick, K. Shanmugam, and I. Dinstein. Textural features for image classification. *IEEE Transactions On Systems, Man, and Cybernetics*, SMC-3(6):610–621, November 1973. DOI: 10.1109/TSMC.1973.4309314. 58

[65] J. Harvda and F. Charvát. Quantification method of classification processes. Concept of structural α-entropy. *Kybernetika*, 3:30–35, 1967. 20

[66] Y. He, A. Hamza, and H. Krim. A generalized divergence measure for robust image registration. *IEEE Transactions on Signal Processing*, 51(5):1211–1220, May 2003. DOI: 10.1109/TSP.2003.810305. 41

[67] E. D. Hellinger. Neue Begründung der Theorie der Quadratischen Formen von Unendlichen Vielen Veränderlichen.[2] *Journal für Reine und Angewandte Mathematik*, 136:210–271, 1909. 20

---

[2]A new foundation of the theory of quadratic forms of infinite many variables.

[68] A. Herbulot, S. Jehan-Besson, M. Barlaud, and G. Aubert. Shape gradient for image segmentation using information theory. In *International Conference on Acoustics, Speech, and Signal Processing*, volume 3, pages 21–24, Montreal, Canada, 2004. DOI: 10.1109/ICASSP.2004.1326471. 63

[69] A. Herbulot, S. Jehan-Besson, S. Duffner, M. Barlaud, and G. Aubert. Segmentation of vectorial image features using shape gradients and information measures. *Journal of Mathematical Imaging and Vision*, 25(3):365–386, 2006. DOI: 10.1007/s10851-006-6898-y. 63, 64

[70] A. O. Hero, B. Ma, O. Michel, and J. Gorman. Applications of entropic spanning graphs. *IEEE Signal Processing Magazine*, 19(5):85–95, September 2002. DOI: 10.1109/MSP.2002.1028355. 35, 37

[71] D. L. Hill. *Combination of 3D medical images from multiple modalities*. PhD thesis, University of Londonn, Image Processing Group, Radiological Sciences, December 1993. 31

[72] F. Hoenig. Defining computational aesthetics. In L. Neumann, M. Sbert, B. Gooch, and W. Purgathofer, editors, *Computational Aesthetics 2005. Eurographics Workshop on Computational Aesthetics in Graphics, Visualization and Imaging*, pages 13–18. Eurographics Association, May 2005. 99

[73] M. Holden. A review of geometric transformations for nonrigid body registration. *IEEE Transactions on Medical Imaging*, 27(1):111–128, 2008. DOI: 10.1109/TMI.2007.904691. 27

[74] M. Holden, L. D. Griffin, and D. L. G. Hill. Multi-dimensional mutual information image similarity metrics based on derivatives of linear scale space. In *Proceedings of the APRS Workshop on Digital Image Computing*, pages 55–60, Brisbane, Australia, February 2005. 39

[75] B. Janvier, E. Bruno, T. Pun, and S. Marchand-Maillet. Information-theoretic temporal segmentation of video and applications: multiscale keyframes selection and shot boundaries detection. *Multimedia Tools and Applications*, 30(3):273–288, 2006. DOI: 10.1007/s11042-006-0026-2. 76, 77

[76] E. T. Jaynes. Information theory and statistical mechanics. *Physical Review*, 106(4):620–630, May 1957. DOI: 10.1103/PhysRev.106.620. 3

[77] A. Journel and C. Deutsch. Entropy and spatial disorder. *Mathematical Geology*, 25(3):329–355, 1993. DOI: 10.1007/BF00901422. 60

[78] G. Jumarie. A new information theoretic approach to the entropy of non-random discrete maps relation to fractional dimension and temperature of curves. *Chaos, Solitons and Fractals*, 8(6):953–970, 1997. DOI: 10.1016/S0960-0779(96)00134-8. 41

[79] A. Kaltchenko. Algorithms for estimating information distance with application to bioinformatics and linguistics. *CoRR*, cs.CC/0404039, April 2004. 43

[80] J. Kapur, P. Sahoo, and A. Wong. A new method for gray-level picture thresholding using the entropy of the histogram. *Computer Vision, Graphics and Image Processing*, 29(3):273–285, March 1985. DOI: 10.1016/S0734-189X(85)90156-2. 54, 59

[81] M. Kass, A. Witkin, and D. Terzopoutlos. Snakes: active countour models. *International Journal Computer Vision*, 1(4):321–331, 1988. DOI: 10.1007/BF00133570. 63

[82] J. Kim, I. John W. Fisher, A. Yezzi, M. Çetin, and A. S. Willsky. A nonparametric statistical method for image segmentation using information theory and curve evolution. *IEEE Transactions on Image Processing*, 14(10):1486–1502, October 2005. DOI: 10.1109/TIP.2005.854442. 63, 64

[83] J. Kittler and J. Illingworth. Minimum error thresholding. *Pattern Recognition*, 19(1):41–47, 1986. DOI: 10.1016/0031-3203(86)90030-0. 56

[84] S. Klein, M. Staring, and J. P. W. Pluim. Evaluation of optimization methods for nonrigid medical image registration using mutual information and b-splines. *IEEE Transactions on Image Processing*, 16(12):2879–2890, December 2007. DOI: 10.1109/TIP.2007.909412. 29

[85] A. N. Kolmogorov. On the Shannon theory of information transmission in the case of continuous signals. *IRE Transactions on Information Theory*, 2:102–108, 1956. DOI: 10.1109/TIT.1956.1056823. 17

[86] M. Koshelev. Towards the use of aesthetics in decision making: Kolmogorov complexity formalizes Birkhoff's idea. *Bulletin of the European Association for Theoretical Computer Science*, 66:166–170, October 1998. 99

[87] A. Kraskov, H. Stögbauer, R. G. Andrzejak, and P. Grassberger. Hierarchical clustering using mutual information. *Europhysics Letters*, 70:278–284, 2005. DOI: 10.1209/epl/i2004-10483-y. 34

[88] S. R. Kulkarni, G. Lugosi, and S. S. Venkatesh. Learning pattern classification – a survey. *IEEE Transactions on Information Theory*, 44(6):2178–2206, 1998. DOI: 10.1109/18.720536. 66

[89] S. Kullback and R. A. Leibler. On information and sufficiency. *Annals of Mathematical Statistics*, 22:76–86, 1951. DOI: 10.1214/aoms/1177729694. 19

[90] S. Lavallee. Registration for computed-integrated-surgery: Methodolgy, state of the art. *Computer Integrated Surgery: Technology and Clinical Applications*, pages 77–97, 1995. 25

[91] F. Lecellier, S. Jehan-Besson, J. Fadili, G. Aubert, and M. Revenu. Optimization of divergences within the exponential family for image segmentation. In *2nd International Conference on Scale Space and Variational Methods in Computer Vision*, June 2009. DOI: 10.1007/978-3-642-02256-2_12. 63, 64

[92] H. C. Lee and S. D. Kim. Rate-driven key frame selection using temporal variation of visual content. *Electronic Letters*, 38(5):217–218, 2002. DOI: 10.1049/el:20020112. 83

[93] T. M. Lehmann, C. Gonner, and K. Spitzer. Registration for computed-integrated-surgery: Methodolgy, state of the art. *IEEE Transactions on Medical Imaging*, 18(11):1049–1074, November 1999. 27

[94] M. Li, J. Badger, X. Chen, S.Kwong, P. Kearney, and H. Zang. An information-based sequence distance and its applications to whole mitochondrial genome phylogeny. *Bioinformatics*, 17(2):149–154, 2001. DOI: 10.1093/bioinformatics/17.2.149. 42

[95] M. Li, X. Chen, X. Li, B. Ma, and P. Vitányi. The similarity metric. *IEEE Transactions on Information Theory*, 50(12):3250–3264, December 2004. DOI: 10.1109/TIT.2004.838101. 22, 23, 34, 42

[96] M. Li and P. Vitányi. *An Introduction to Kolmogorov Complexity and Its Applications*. Graduate Texts in Computer Science. Springer-Verlag, New York (NY), USA, 1997. 22, 42

[97] M. Li and P. M. B. Vitányi. *An Introduction to Kolmogorov Complexity and Its Applications*. Graduate Texts in Computer Science. Springer-Verlag, 1997. 102

[98] R. Lienhart. Reliable transition detection in videos: A survey and practitioner's guide. *International Journal of Image and Graphics*, 1:469–486, 2001. DOI: 10.1142/S021946780100027X. 75

[99] R. Lienhart, S. Pfeiffer, and W. Effelsberg. Video abstracting. *Commun ACM*, 40(12):54–62, December 1997. DOI: 10.1145/265563.265572. 75, 78

[100] T.-C. Liu and J. R. Kender. Computational approaches to temporal sampling of video sequences. *ACM Transactions on Multimedia Computing*, 3(2):217–218, 2007. DOI: 10.1145/1230812.1230813. 84

[101] A. J. Lubin. *Stranger On The Earth: A Psychological Biography Of Vincent Van Gogh*. Da Capo Press, 1996. 110

[102] B. Ma, A. Hero, J. Gorman, and O. Michel. Image registration with minimum spanning tree algorithm. In *International Conference on Image Processing, 2000*, volume 1, pages 481–484, 2000. DOI: 10.1109/ICIP.2000.901000. 37

[103] P. Machado and A. Cardoso. Computing aesthetics. In *Proceedings of XIVth Brazilian Symposium on Artificial Intelligence (SBIA '98)*, LNAI, pages 219–229, Porto Alegre, Brazil, November 1998. Springer-Verlag. DOI: 10.1007/10692710_23. 99

[104] F. Maes, A. Collignon, D. Vandermeulen, G. Marchal, and P. Suetens. Multimodality image registration by maximization of mutual information. *IEEE Transactions on Medical Imaging*, 16(2):187–198, 1997. DOI: 10.1109/42.563664. 27, 29, 32, 33, 34, 35, 46

[105] F. Maes, D. Vandermeulen, and P. Suetens. Comparative evaluation of multiresolution optimization strategies for multimodality image registration by maximization of mutual information. *Medical Image Analysis*, 3(4):373–386, 1999. DOI: 10.1016/S1361-8415(99)80030-9. 29

[106] D. Martin, C. Fowlkes, D. Tal, and J. Malik. A database of human segmented natural images and its application to evaluating segmentation algorithms and measuring ecological statistics. In *Proc. 8th Int'l Conf. Computer Vision*, volume 2, pages 416–423, July 2001. DOI: 10.1109/ICCV.2001.937655. 55

[107] A. F. T. Martins, M. A. T. Figueiredo, P. M. Q. Aguiar, N. A. Smith, and E. P. Xing. Nonextensive entropic kernels. In *Proceedings of the 25th International Conference on Machine Learning*, ICML '08, pages 640–647, 2008. DOI: 10.1145/1390156.1390237. 81

[108] D. Mattes, D. R. Haynor, H. Vesselle, T. K. Lewellen, and W. Eubank. Non-rigid multimodality image registration. In *SPIE Medical Imaging 2001: Image Processing*, volume 4322, pages 1609–1620, 2001. 36

[109] M. Mentzelopoulos and A. Psarrou. Key-frame extraction algorithm using entropy difference. In *Proc. ACM SIGMM Int. Conf. Workshop Multimedia Information Retrieval*, pages 39–45, 2004. DOI: 10.1145/1026711.1026719. 76

[110] A. Moles. *Information Theory and Esthetic Perception*. University of Illinois Press, 1968. 98, 99

[111] Piet Mondrian. Composition No. 1 with Grey and Red 1938 / Composition with Red 1939, 1938–39. The Solomon R. Guggenheim Foundation, Peggy Guggenheim Collection. http://www.guggenheim.org/new-york/collections/collection-online/artwork/3053 100

[112] Piet Mondrian. Composition with Red, Blue, Black, Yellow, and Gray, 1921. The Museum of Modern Art. http://www.moma.org/collection/object.php?object_id=79002 100

[113] Piet Mondrian. Composition with Grid 1, 1918. Museum of Fine Arts. http://www.mfah.org/art/100-highlights/composition-grid-1/ 100

[114] A. G. Money and H. Agius. Video summarisation: A conceptual framework and survey of the state of the art. *Journal of Visual Communication and Image Representation*, 19(2):121–143, February 2008. DOI: 10.1016/j.jvcir.2007.04.002. 76

[115] S. Naifeh and G. W. Smith. *Van Gogh: The Life*. Random House, 2012. 110

[116] F. Nake. *Ästhetik als Informationsverarbeitung: Grundlagen und Anwendungen der Informatik im Bereich ästhetischer Produktion und Kritik (Asthetics as Data Processing: Bases and Applications of Computer Science in the Area of Aesthetic Production and Criticism)*. Springer-Verlag, 1974. 103

[117] M. Omidyeganeh, S. Ghaemmaghami, and S. Shirmohammadi. Video keyframe analysis using a segment-based statistical metric in a visually sensitive parametric space. *IEEE Transactions of Image Processing*, 20(10):2730–2737, October 2011. DOI: 10.1109/TIP.2011.2143421. 76, 77

[118] S. Osher and J. A. Sethian. Fronts propagation with curvature-dependent speed: algorithms based on hamilton-jacobi formulations. *Journal of Computiacional Physics*, 79:12–49, 1988. DOI: 10.1016/0021-9991(88)90002-2. 63

[119] J. A. O'Sullivan, R. E. Blahut, and D. L. Snyder. Information-Theoretic Image Theory, *IEEE Transactions on Information Theory*, 44(6):2094–2123, 1998. xi

[120] N. Pal and S. Pal. Object-background segmentation using new definitions of entropy. *Computers and Digital Techniques, IEE Proceedings*, 136:284–295, July 1989. DOI: 10.1049/ip-e.1989.0039. 58, 59

[121] N. Pal and S. Pal. Image model, Poisson distribution and object extraction. *International Journal of Pattern Recognition and Artificial Intelligence*, 5(3):459–483, 1991. DOI: 10.1142/S0218001491000260. 57

[122] A. Papoulis. *Probability, Random Variables, and Stochastic Processes*. McGraw-Hill, New York (NY), USA, 1984. 3

[123] M. Pardo and I. Vajda. On asymptotic properties of information-theoretic divergences. *IEEE Transactions on Information Theory*, 49(7):1860–1868, 2003. DOI: 10.1109/TIT.2003.813509. 18

[124] E. Parzen. On estimation of a probability density function and mode. *Annals of Mathematical Statistics*, 33(3):1065–1076, 1962. DOI: 10.1214/aoms/1177704472. 35, 36

[125] K. Pearson. On the criterion that a given system of deviations from the probable in the case of a correlated system of variables is such that it can be reasonably supposed to have arisen from random sampling. *Philosophical Magazine*, V(1):157–175, 1900. DOI: 10.1080/14786440009463897. 20

[126] D. L. Pham, C. Xu, and J. L. Prince. A survey of current methods in medical image segmentation. Technical Report JHU/ECE 99-01, Department of Electrical and Computer Engineering, Johns Hopkins University, Baltimore (MD), USA, 1998. 53

[127] R. Pickvance. *Van Gogh in Saint-Rémy and Auvers*. The Metropolitan Museum of Art (Harry N. Abrams, Inc.), 1986. Published in conjunction with the exhibition *Van Gogh in Saint-Rémy and Auvers*, held at The Metropolitan Museum of Art, New York, 25 November 1986 – 22 March 1987. 117, 118, 120

[128] M. S. Pinsker. *Information and Stability of Random Variables and Processes*. Izdatel'stvo Akademii Nauk SSSR, Moscow, Russia, 1960. 17

[129] J. P. Pluim. *Mutual Information Based Registration of Medical Images*. PhD thesis, Image Sciences Institute, Utrecht, The Netherlands, 2001. 25

[130] J. P. Pluim, J. Maintz, and M. Viergever. Image registration by maximization of combined mutual information and gradient information. *IEEE Transactions on Medical Imaging*, 19(8):809–814, 2000. DOI: 10.1109/42.876307. 38

[131] J. P. Pluim, J. Maintz, and M. Viergever. Interpolation artifacts in mutual information based image registration. In *Computer Vision and Image Understanding*, volume 77, pages 211–232, 2000. DOI: 10.1006/cviu.1999.0816. 28

[132] J. P. Pluim, J. Maintz, and M. Viergever. Mutual-information-based registration of medical images: a survey. *IEEE Transactions on Medical Imaging*, 22:986–1004, 2003. DOI: 10.1109/TMI.2003.815867. 29, 30, 32, 36, 43

[133] J. P. Pluim, J. Maintz, and M. Viergever. f-information measures in medical image registration. *IEEE Transactions on Medical Imaging*, 23(12):1508–1516, December 2004. DOI: 10.1109/TMI.2004.836872. 39, 40

[134] M. Portes de Albuquerque, I. Esquef, A. G. Mello, and M. Portes de Albuquerque. Image thresholding using Tsallis entropy. *Pattern Recognition Letters*, 25:1059–1065, 2004. DOI: 10.1016/j.patrec.2004.03.003. 55, 80

[135] W. Press, S. Teulokolsky, W. Vetterling, and B. Flannery. *Numerical Recipes in C*. Cambridge University Press, 1992. 29, 31

[136] T. Pun. A new method for gray-level picture thresholding using the entropy of the histogram. *Signal Processing*, 2:223–237, 1980. DOI: 10.1016/0165-1684(80)90020-1. 54

[137] T. Pun. Entropic thresholding: a new approach. *Computer Graphics and Image Processing*, 16:210–239, 1981. DOI: 10.1016/0146-664X(81)90038-1. 54

[138] A. Rényi. On measures of entropy and information. In *Proc. Fourth Berkeley Symp. Math. Stat. and Probability' 60*, volume 1, pages 547–561, Berkeley (CA), USA, 1961. University of California Press. 20

[139] J. Rigau, M. Feixas, and M. Sbert. An information theoretic framework for image segmentation. In *IEEE International Conference on Image Processing (ICIP '04)*, volume 2, pages 1193–1196, Victoria (British Columbia), Canada, October 2004. IEEE Press. DOI: 10.1109/ICIP.2004.1419518. 65, 66, 69, 108

[140] J. Rigau, M. Feixas, and M. Sbert. Conceptualizing Birkhoff's aesthetic measure using Shannon entropy and Kolmogorov complexity. In D. W. Cunningham, G. Meyer, L. Neumann, A. Dunning, and R. Paricio, editors, *Computational Aesthetics 2007. Eurographics Workshop on Computational Aesthetics in Graphics, Visualization and Imaging*, pages 105–112. Eurographics Association, June 2007. 97

[141] J. Rigau, M. Feixas, and M. Sbert. Informational aesthetics measures. *IEEE Computer Graphics and Applications*, 28(2):24–34, March/April 2008. DOI: 10.1109/MCG.2008.34. 66, 99, 102, 105, 107, 108, 110, 111, 115

[142] J. Rigau, M. Feixas, and M. Sbert. Informational dialogue with Van Gogh's paintings. In P. Brown, D. W. Cunningham, V. Interrante, and J. McCormack, editors, *Computational Aesthetics 2008. Eurographics Workshop on Computational Aesthetics in Graphics, Visualization and Imaging*, pages 115–122. Eurographics Association, June 2008. 66, 97, 99, 110, 113, 115, 122

[143] J. Rigau, M. Feixas, M. Sbert, and C. Wallraven. Towards Auvers period: Evolution of van Gogh's style. In O. Deussen and P. Jepp, editors, *Computational Aesthetics 2010. Eurographics Workshop on Computational Aesthetics in Graphics, Visualization and Imaging*, pages 99–106. Eurographics Association, June 2010. 98, 99, 119, 123, 124

[144] C. E. Rodríguez-Carranza and M. H. Loew. A weighted and deterministic entropy measure for image registration using mutual information. In *Proceedings of Medical Imaging SPIE 1998*, volume 3338, pages 155–166, San Diego, USA, 1998. 41, 42

[145] D. Rueckert, M. J. Clarkson, D. L. J. Hill, and D. J. Hawkes. Non-rigid registration using higher order mutual information. In *Medical Imaging: Image Processing*, pages 438 – 447. Ed. Bellingham, SPIE Press, 2000. 38

[146] D. B. Russakoff, C. Tomassi, T. Rohlfing, and C. R. Maurer. Image similarity using mutual information of regions. In *8th European Conference on Computer Vision*, volume 3, pages 596 – 608, Prague, Czech Republic, May 2004. DOI: 10.1007/978-3-540-24672-5_47. 39

[147] M. R. Sabuncu and P. J. Ramadge. Spatial information in entropy-based image registration. In *Proceedings of 2nd Workshop in Biomedical Image Registration (WBIR'03)*, volume 2717 of *Lecture Notes in Computer Science*, pages 132–141. Philadelphia, USA, May 2003. DOI: 10.1007/978-3-540-39701-4_14. 39

[148] P. Sahoo, C. Wilkins, and J. Yeager. Threshold selection using Renyi's entropy. *Pattern Recognition*, 30(1):71–84, January 1997. DOI: 10.1016/S0031-3203(96)00065-9. 55

[149] P. K. Sahoo, S. Soltani, A. K. Wong, and Y. C. Chen. A survey of thresholding techniques. *Comput. Vision Graph. Image Process.*, 41(2):233–260, 1988. DOI: 10.1016/0734-189X(88)90022-9. 54

[150] O. Salvado and D. Wilson. Removal of interpolation induced artifacts in similarity surfaces. In *Proceedings of the Third International Workshop on Biomedical Image Registration (WBIR 2006)*, volume 4057 of *Lecture Notes in Computer Science*, pages 43–49. Springer Berlin / Heidelberg, Utrecht, Netherlands, 2006. DOI: 10.1007/11784012_6. 28

[151] L. A. Santaló. *Integral Geometry and Geometric Probability*. Cambridge University Press, 1976. 38

[152] M. Sbert. An integral geometry based method for fast form-factor computation. *Computer Graphics Forum (Proceedings of Eurographics '93)*, 12(3):409–420, 1993. DOI: 10.1111/1467-8659.1230409. 38, 62

[153] M. Sbert, M. Feixas, J. Rigau, M. Chover, and I. Viola. *Information Theory Tools for Computer Graphics*, Morgan & Claypool Publishers, 2009. xi

[154] R. Scha and R. Bod. Computationele esthetica. *Informatie en Informatiebeleid*, 11(1):54–63, 1993. 97, 98

[155] I. K. Sethi and G. Sarvarayudu. Hierarchical classifier design using mutual information. *IEEE Transactions on Pattern Analysis and Machine Intelligence*, 4(4):441–445, July 1982. DOI: 10.1109/TPAMI.1982.4767278. 66, 67

[156] M. Sezgin and B. Sankur. Survey over image thresholding techniques and quantitative performance evaluation. *Journal of Electronic Imaging*, 13(1):146–168, January 2004. DOI: 10.1117/1.1631315. 54

[157] C. E. Shannon. A mathematical theory of communication. *The Bell System Technical Journal*, 27:379–423, 623–656, July, October 1948. DOI: 10.1002/j.1538-7305.1948.tb00917.x. 1, 2, 3

[158] B. D. Sharma and D. P. Mittal. New non-additive measures of entropy for a discrete probability distribution. *Journal of Mathematical Sciences (India)*, 10:28–40, 1975. 20

[159] B. D. Sharma and I. J. Taneja. Entropy of type (α, β) and other generalized measures in information theory. *Metrika*, 22(1):205–215, 1975. DOI: 10.1007/BF01899728. 20

[160] N. Slonim and N. Tishby. Agglomerative information bottleneck. In *Proceedings of NIPS-12 (Neural Information Processing Systems)*, pages 617–623. MIT Press, 2000. 18, 67, 71

[161] N. Slonim and N. Tishby. Document clustering using word clusters via the information bottleneck method. In *Proceedings of the 23rd Annual International ACM SIGIR Conference on Research and Development in Information Retrieval*, pages 208–215. ACM Press, 2000. DOI: 10.1145/345508.345578. 12

[162] G. Solana. Epilogue in Auvers. In *Van Gogh. Los Últimos Paisajes*, pages 153–163. Fundación Colección Thyssen-Bornemisza, 2007. 118

[163] M. A. Stricker and M. Orengo. Similarity of color images. *Proc SPIE*, 2420(2):381–392, May 1995. DOI: 10.1117/12.205308. 84

[164] C. Studholme. *Measures of 3D medical image alignment*. PhD thesis, Computational Imaging Science Group, Division of Radiological Sciences, United Medical and Dental school's of Guy's and St Thomas's Hospitals, 1997. 29, 34

[165] C. Studholme, D. Hill, and D. Hawkes. Multiresolution voxel similarity measures for mr-pet registration. In *Proceedings of the Information Processing in Medical Imaging Conference*, pages 287–298, Ile de Berder, France, June 1995. 31, 33

[166] C. Studholme, D. Hill, and D. Hawkes. An overlap invariant entropy measure of 3D medical image alignment. *Pattern Recognition*, 32(1):71–86, 1999. DOI: 10.1016/S0031-3203(98)00091-0. 30, 34

[167] I. J. Taneja. Bivariate measures of type α and their applications. *Tamkang Journal of Mathematics*, 19(3):63–74, 1988. 21

[168] W. B. Thompson, R. W. Fleming, S. H. Creem-Regehr, and J. K. Stefanucci. *Visual Perception from a Computer Graphics Perspective*. CRC Press, 2011. 80

[169] N. Tishby, F. C. Pereira, and W. Bialek. The information bottleneck method. In *Proceedings of the 37th Annual Allerton Conference on Communication, Control and Computing*, pages 368–377, 1999. 17, 18

[170] B. T. Truong and S. Venkatesh. Video abstraction: A systematic review and classification. *ACM T. Multim. Comput.*, 3(1):1–37, 2007. DOI: 10.1145/1198302.1198305. 76, 83

[171] C. Tsallis. Possible generalization of Boltzmann-Gibbs statistics. *Journal of Statistical Physics*, 52(1/2):479–487, 1988. DOI: 10.1007/BF01016429. 21

[172] C. Tsallis. Generalized entropy-based criterion for consistent testing. *Physical Review E*, 58:1442–1445, 1998. DOI: 10.1103/PhysRevE.58.1442. 21

[173] C. Tsallis. Entropic nonextensivity: A possible measure of complexity. *Chaos, Solitons, & Fractals*, 13(3):371–391, 2002. DOI: 10.1016/S0960-0779(01)00019-4. 21

[174] J. Tsao. Interpolation artifacts in multimodal image registration based on maximization of mutual information. *IEEE Transactions on Medical Imaging*, 22:854–864, November 2003. DOI: 10.1109/TMI.2003.817660. 28

[175] M. Unser. Splines. A perfect fit for signal and image processing. *IEEE Signal Processing Magazine*, 16:22–38, 1999. DOI: 10.1109/79.799930. 27

[176] M. Unser and P. Thévenaz. Stochastic sampling for computing the mutual information of two images. In *5th International Workshop on Sampling Theory and Applications, 2003, Proceedings.*, pages 102–109, Strobl, Austria, May 2003. 28

[177] S. Verdú. Fifty years of Shannon theory. *IEEE Transactions on Information Theory*, 44(6):2057–2078, October 1998. DOI: 10.1109/18.720531. 1

[178] M. Vila, A. Bardera, Q. Xu, M. Feixas, and M. Sbert. Tsallis entropy-based information measures for shot boundary detection and keyframe selection. *Signal, Image and Video Processing*, 7(3):507–520, 2013. DOI: 10.1007/s11760-013-0452-3. 77, 80, 84, 90, 94, 95

[179] P. A. Viola. *Alignment by Maximization of Mutual Information*. PhD thesis, MIT Artificial Intelligence Laboratory (TR 1548), Massachusetts (MA), USA, 1995. 29, 35, 36, 37, 46

[180] M. P. Wachowiak, R. Smolikova, G. D. Tourassi, and A. S. Elmaghraby. Similarity metrics based on non-additive entropies for 2D-3D multimodal biomedical image registration. In *Proceedings of SPIE Medical Imaging'03*, pages 1090–1100, 2003. DOI: 10.1117/12.480867. 40, 41

[181] C. Wallraven, D. Cunningham, J. Rigau, M. Feixas, and M. Sbert. Aesthetic appraisal of art — from eye movements to computers. In O. Deussen, P. Hall, S. Gibson, G. Hushlack, and J. Shaw, editors, *Computational Aesthetics 2009. Eurographics Workshop on Computational Aesthetics in Graphics, Visualization and Imaging*, pages 137–144. Eurographics Association, May 2009. 115

[182] C. Wallraven, R. Fleming, D. Cunningham, J. Rigau, M. Feixas, and M. Sbert. Categorizing art: Comparing humans and computers. *Computers & Graphics*, 33(4):484–495, August 2009. DOI: 10.1016/j.cag.2009.04.003. 115

[183] Q. Wang, S. R. Kulkarni, and S. Verdú. *Universal Estimation of Information Measures for Analog Sources*. Foundations and Trends in Communications and Information Theory: Vol. 5: No 3, pp 265-353. Now Publishers, 2009. 37

[184] W. Wells, P. Viola, H. Atsumi, S. Nakajima, and R. Kikinis. Multi-modal volume registration by maximization of mutual information, 1996. DOI: 10.1016/S1361-8415(01)80004-9. 32

[185] B. Wilson, E. B. Lum, and K.-L. Ma. Interactive multi-volume visualization. In *Proceedings of the International Conference on Computational Science-Part II*, pages 102–110, 2002. DOI: 10.1007/3-540-46080-2_11. 44

[186] Q. Xu, X. Li, Z. Yang, J. Wang, M. Sbert, and J. Li. Key frame selection based on Jensen-Rényi divergence. In *Pattern Recognition (ICPR), 2012 21st International Conference on*, pages 1892–1895, 2012. 77, 79, 88, 89, 90, 91, 92, 94

[187] Q. Xu, P. Wang, B. Long, M. Sbert, M. Feixas, and R. Scopigno. Selection and 3D visualization of video key frames. In *Systems Man and Cybernetics (SMC), 2010 IEEE International Conference on*, pages 52–59, 2010. DOI: 10.1109/ICSMC.2010.5642204. 77, 85, 86, 94

[188] R. W. Yeung. *Information Theory and Network Coding*. Information Technology: Transmission, Processing and Storage. Springer, 2008. 1, 3, 7, 10, 13, 15

[189] J. Yuan, H. Wang, L. Xiao, W. Zheng, J. Li, F. Lin, and B. Zhang. A formal study of shot boundary detection. *IEEE Transactions on Circuits and Systems for Video Technology*, 17(2):168 –186, February 2007. DOI: 10.1109/TCSVT.2006.888023. 75

[190] H. Yun, A. B. Hamza, and H. Krim. A generalized divergence measure for robust image registration. *IEEE Trans Signal Process*, 51(5):1211–1220, May 2003. DOI: 10.1109/TSP.2003.810305. 78

[191] W. Zhang, J. Lin, X. Chen, Q. Huang, and Y. Liu. Video shot detection using hidden markov models with complementary features. In *First International Conference on Innovative Computing, Information and Control, ICICIC '06*, volume 3, pages 593 –596, September 2006. DOI: 10.1109/ICICIC.2006.549. 77

[192] W. H. Zurek. Algorithmic randomness and physical entropy. *Physical Review D*, 40(8):4731–4751, 1989. DOI: 10.1103/PhysRevA.40.4731. 103

Authors' Biographies

MIQUEL FEIXAS

Miquel Feixas is an associate professor in Computer Science at the University of Girona. He received an M.Sc. in Theoretical Physics (1979) at the Universitat Autònoma de Barcelona and a Ph.D. in Computer Science (2002) at the Universitat Politècnica de Catalunya. His research is focused on the application of Information Theory techniques to Radiosity, Global Illumination, Viewpoint Selection, Scientific Visualization, Image Processing, Medical Imaging, and Computational Aesthetics. He has co-authored more than eighty papers in his areas of research. He has participated in several European and Spanish research projects.

ANTON BARDERA

Anton Bardera is a lecturer in Computer Science at the University of Girona. He received a M.Sc. in Telecommunications engineering (2002) at the Universitat Politècnica de Catalunya and a Ph.D. in Computer Science (2008) at the University of Girona. His research interests include Image Processing, Information Theory, and Biomedical Applications. He has authored or co-authored more than thirty papers in his areas of interest.

JAUME RIGAU

Jaume Rigau is an associate professor in Computer Science at the University of Girona. He received an M.Sc. in Computer Science (1993) and a Ph.D. (2006) at the Universitat Politècnica de Catalunya. His research is focused on the application of Information Theory to Computer Graphics and Image Processing. He has co-authored more than thirty papers in his areas of research and participated in European and Spanish research projects.

QING XU

Qing Xu is currently a full professor with the School of Computer Science and Technology, Tianjin University, China, where he is the director of the Information Technology Research and Civil Aviation Applications Laboratory. He got a M.Sc. (1998) and a Ph.D. (2001) in Computer Science at Tianjin University. His research activities cover interdisciplinary fields related to Computer Graphics, Visualization and Visual Analytics, Digital Image/Video Processing, Information Theory and Fuzzy Techniques. He has been leading national, international cooperation and industrial research projects. The results of his research have result in sixty publications. He also serves for the program committee and/or reviewer of a number of conferences/journals in computer graphics, virtual realities, and image/video processing.

MATEU SBERT

Mateu Sbert is a full professor in Computer Science at the University of Girona, Spain. He received an M.Sc. in Theoretical Physics (1977) at the University of Valencia, an M.Sc. in Mathematics (Statistics and Operations Research, 1983) at U.N.E.D. University (Madrid), and a Ph.D. in Computer Science (1997, Best Ph.D. Award) at the Universitat Politècnica de Catalunya. Mateu Sbert's research interests include the application of Monte Carlo and Information Theory techniques to Computer Graphics and Image Processing. He has authored or co-authored around two hundred papers in his areas of research, served as a member of program committee in Spanish and international conferences, and led European and Spanish research projects. Mateu Sbert coorganized the Dagsthul Seminars "Stochastic Methods in Rendering" and "Computational Aesthetics in Graphics, Visualization and Imaging."

Printed in the United States
by Baker & Taylor Publisher Services